Welcome

to *Triumph of Discovery*, a celebration of 150 years of *Scientific American*. In the pages that follow, you will find 48 essays on scientific explorations that have intrigued the human imagination for at least the last century and a half. Others address more recent investigations now in today's headlines.

Each essay, illustrated with recent and vintage images, is written by a leading scholar, specifically for this commemorative work.

Also, throughout the book, you will find chronologies—"snapshots" capturing the mood of each decade, highlighting key events in science, technology, medicine, politics and popular culture since the time when *Scientific American* began publication in 1845.

The publisher and authors of *Triumph of Discovery* hope you enjoy this remarkable presentation—celebrating not only the triumphs of one magazine, but of our individual and collective scientific achievements.

SCIENTIFIC AMERICAN

Triumph
of
Discovery

A Chronicle
of Great Adventures
in Science

FOREWORD BY JOHN H. GIBBONS

*Science Advisor to the
President of the United States*

A HENRY HOLT REFERENCE BOOK

HENRY HOLT AND COMPANY

NEW YORK

A Henry Holt Reference Book
Henry Holt and Company, Inc.
Publishers since 1866
115 West 18th Street
New York, New York 10011

Published in Canada by Fitzhenry & Whiteside Ltd.,
195 Allstate Parkway, Markham, Ontario L3R 4T8.

Library of Congress Cataloging-in-Publication Data
Scientific American : triumph of discovery : a chronicle of great adventures in science /
foreword by John H. Gibbons. — 1st ed.
p. cm — (Henry Holt reference book)
Includes index.
1. Science. 2. Scientific American. I. Scientific American, Inc.
II. Title: Triumph of discovery. III. Series.
Q158.5.S356 1995   94-39516
500—dc20   CIP
ISBN 0-8050-3551-6

Henry Holt books are available for special promotions and premiums.
For details contact: Director, Special Markets.

First Edition—1995

Conceived by Henry Holt and Company, Inc. and Robert Ubell Associates, Inc.
Project Director: Robert N. Ubell
Project Manager: Luis A. Gonzalez
Design: Foca Company
Design Director: Melanie Roberts

Printed in the United States of America

All first editions are printed on acid-free paper.∞
1 3 5 7 9 10 8 6 4 2

A BOOK HOLDER,

ILLUSTRATED IN AN 1884 ISSUE

OF *SCIENTIFIC AMERICAN*

Contents

Essays

CONTENTS

Chronologies

Scientific American

AN ILLUSTRATED

JOURNAL ART. SCIENCE & MECHANICS

Vol. XXI.

NEW-YORK

PUBLISHED BY MUNN & CO.

Foreword

Most of us spend part of each day in complete awe of science and technology — or would if there were time. Some things we take for granted, like cold drinks in the refrigerator or rapid air and ground transportation. Many things still seem fantastic, like satellite photography or a child joyfully conceived because of infertility treatments. Whether oblivious or grateful, scientists or laymen, we are all beneficiaries of a Jeffersonian revolution—"the extraordinary event necessary to enable all the ordinary events to continue."

Tuesday, 6:45 AM. This is my first chance to prepare tonight's speech at the Air and Space Museum. Comparing the first photograph of the sun, taken in 1855, with the latest, show-stopping revelations of the Hubble Telescope, I marvel at the pace of discovery. We know the distances to over 100 billion stars, yet at the turn of the century we had thoroughly charted only one hundred.

The universe has stimulated immense curiosity and scientific inquiry since people learned how to ask questions. Stargazing affords pleasure to all, and more advanced study of the universe increases our storehouse of knowledge. Individuals and countries bold enough to ask, "Does the earth revolve around the sun?" or "Can we safely visit the moon?" have radically altered our lives.

Outer space has long gripped our collective imagination, and exploration of "inner space" represents another endless frontier. Employing advanced microscopes and computer simulations, we can examine the atomic structure of materials. Scientists can determine, for instance, how a single oxygen molecule upsets a silicon lattice. Research that promises improved manufacturing processes for semiconductor chips will lead to faster, cheaper computers. We can also look deeply into the molecular structure of living cells. Undertakings like the Human Genome Project will not only produce a catalogue of human genes, but will ensure an ever-increasing supply of knowledge and generate enthusiastic young scientists who will ask the right questions.

Tuesday, 9:15 AM. On the way to a meeting on the Federal research and development budget, I have time to reflect on this morning's telecommunications breakfast with the Vice President. As we talked about the possibilities for linking libraries and other educational centers in this hemisphere through a global information network, we were interrupted by urgent "faxes" and telephone calls from around the world—a telling conjuncture of technology's power to enable and to irritate.

Readers of the first issue of *Scientific American* would marvel at our telephones, radios and television sets. We now soothe our children with tapes on the VCR, quickly adopted the fax machine as essential office equipment, and appear ready to accelerate onto the information superhighway.

The speed and power of modern information technologies enable diverse activities, from office work, to constructing computer models of the likely impacts of climate change, to probing atomic and molecular structures.

Technologies like the Global Positioning System (GPS), which uses satellites to pinpoint location, make the lives of the Jetsons seem almost real, even as we apply them to common activities such as emergency rescue or tracking commercial vehicles. The U.S. military created GPS for defense purposes, but now more than 160 manufacturers are introducing a variety of GPS-based systems worldwide for a new multibillion dollar commercial market.

The U.S. government has a long-standing commitment to funding scientific research, the results of which are often transferred to the private sector. James Madison said, "A people who mean to be their own governors must have the power that knowledge gives." Science and technology are fundamental to our knowledge-based society.

Tuesday, 10:35 AM. Next on the agenda is a meeting on biotechnology with members of the congressional appropriations committees. The diversity of applications for biotechnology—health care, food production, pollution control—complicates the effort to decide on national policy. Biotechnology will eventually touch the lives of most people around the world, so it's essential to understand how Federal research dollars and regulatory programs affect private sector efforts.

Biotechnology has been with us almost as long as astronomy. People have made beer, cheese, and leavened bread for at least 6000 years. Yet it wasn't until the mid-nineteenth century that Louis Pasteur's experiments proved that the technology that produces those foods—fermentation—results from a biological process. In recent decades, fermentation has been joined by a number of very sophisticated bioprocesses, systems in which living cells or their components are used to create physical or chemical changes. Bioremediation, a biological conversion process in which living organisms decompose or store waste byproducts, is being used to treat contaminated and radioactive wastes, oil spills, acid wastes, municipal wastes, and pesticides.

Biotechnologies will also be used to produce safer, more nutritious food, to create crops that can adapt to climate change and resist pests, and to contribute to farming techniques that conserve or reclaim fragile soils. The agriculture industry accounts for 15 percent of the U.S. gross domestic product, rings up foreign sales of $40 billion annually, and employs 21 million Americans. Biotechnology will bring certain change to the way farmers do business, and it will very likely sustain this vital industry well into the future.

The pace of technological advance, in biotechnology or any research arena, often outstrips our abilities—as individuals and as a society—to accommodate change. When Henry Adams first encountered the dynamo and the profound implications of electric power, he found himself in the Gallery of Machines at the Great Exposition of 1900, "his historical neck broken by the sudden irruption of forces totally new." Today, we cannot imagine life without electricity, it will soon be true of biotechnology.

Tuesday, 12:20 PM. A phone call from my doctor confirms that biotechnology saved my life. A new diagnostic test that first revealed a pre-symptomatic cancer now shows, after treatment, that I'm cancer-free. Information technology has saved my neck. Later this afternoon, I'll videotape my speech so I can make it to my grandson's birthday dinner scheduled for the same time. In the meantime, the afternoon is chock-full—a meeting with other Presidential advisors on nuclear nonproliferation, a conference call with auto-industry executives to discuss new technologies for safe, low polluting-cars, and briefings from my staff on high-energy physics and biodiversity.

Science and technology give us enormous power. We not only depend on them for the basic things that occupy our day—jobs, transportation, education—but we also count on science and technology for giving us a sense of progress and intellectual satisfaction.

John H. Gibbons
Science Advisor to the President of the United States

VIEW OF THE *SCIENTIFIC AMERICAN* OFFICE IN NEW YORK,

ILLUSTRATED IN AN 1859 ISSUE

Introduction

150 years seems to most of us a very long time, and no other American magazine has published continuously for so long. But 150 centuries better describes the duration of the engagement of human beings with science. Dwellers at the foot of the receding Asian glacier, thousands of years ago, lacked fuel, and so used trial and error to learn to kindle wooly-mammoth bones. Those who acquired the knowledge were revered and every tribe needed to hear and share it. Today, we store and communicate lore in books, newspapers, journals, computers, and in *Scientific American* (now available online). Tomorrow we may be publishing an edition for space travelers colonizing Titan.

Just before *Scientific American*'s centennial year, a young man from Harvard, Gerard Piel (who still claims he flunked high-school physics), was editing science for *Life* magazine. Using what he called "little words and big pictures," Piel fostered a collaboration between scientist, editor, and artist. He also met a colleague, Dennis Flanagan, who confidently predicted, "What this country needs is a good magazine of science." In fact, America had a once-proud magazine of science which had been brought into existence in 1845 by an extraordinarily creative individual, Rufus Porter. But by 1947, the publication was much reduced in circumstance and Piel arranged to buy it. Flanagan became its editor.

It was an exciting time. The Second World War had stimulated major advances in science, technology, and medicine. A growing number of the scientifically learned began to speak out about their discoveries. Piel, Flanagan and others filled *Scientific American* with these astonishing stories. Since the re-founding of the publication, more than 100 Nobel laureates have written articles, often prior to the recognition of their work. Editions in German, Italian, French, Spanish, Japanese and several other languages are purchased and read monthly by nearly a million inhabitants of our scientifically connected "global village." At its sesquicentennary, *Scientific American* is still growing, reaching further and further into all the branches of science and medicine, increasingly fulfilling its educational duty, lending its name and guidance to an award-winning series on public television, and employing hundreds of committed people in various sectors of publishing, printing, and distribution. Its ownership and perspective are literally international, unfailingly enthusiastic and optimistic about the promises of science, and determined to fill the need for its communication at the highest level of quality.

On behalf of our many colleagues who have labored with such dedication to make *Scientific American* the world's standard in science publishing, it is a pleasure to thank the hundreds of authors and artists who have made our efforts possible, the millions of subscribers and newsstand buyers who have provided the magazine's basic sustenance, the advertisers who have financed its growth and development, and the leaders who kept the enterprise on course: original founder, Rufus Porter; neo-founder, Gerard Piel; and the current chairman of our publishing organization, Dieter von Holtzbrinck. The excitement hasn't faded. Imagine what we will publish over the next 150 years!

John J. Moeling, Jr.

President and Publisher, Scientific American, Inc.

SCIENTIFIC AMERICAN

Triumph
of
Discovery

1845

SCIENTIFIC AMERICAN is founded in New York City.

▲

BRITISH PHYSICIST JAMES JOULE discovers the law of conservation of energy. The law, also known as the first law of thermodynamics, states that the total amount of energy in the universe is constant; none can be created and none destroyed. It is generally viewed as the most basic of all the laws of nature.

46

MASSACHUSETTS DENTIST WILLIAM MORTON uses ether on a patient while extracting a tooth. The patient tells the local paper, and Morton is urged to perform a demonstration at Massachusetts General Hospital. The demonstration introduces the practice of anesthesia into medicine.

47

ITALIAN CHEMIST ASCANIO SOBRERO mixes nitric and sulfuric acids with glycerin, producing nitroglycerin. When he heats a single drop in a test tube, it explodes with such force that he immediately stops his research. Still, the era of modern explosives has begun.

MUNN & COMPANY purchases *Scientific American* from Rufus Porter.

48

KARL MARX AND FRIEDRICH ENGELS, German socialist philosophers, write *The Communist Manifesto*.

CHEMISTRY OF DESIRE

What is addiction? Despite the vast social impact of addicting drugs, full consensus on a definition has yet to emerge, although authorities do agree on what the key properties of an addictive substance are. One becomes tolerant to an addictive agent so that with chronic use, higher doses must be used to obtain an effect. But tolerance occurs to many nonaddictive drugs, such as penicillin. A more important criterion is physical dependence, which refers to the withdrawal symptoms that occur when one abruptly terminates chronic use of a drug. When heroin addicts stop taking the

Solomon H. Snyder

drug, they become extremely anxious and shiver extensively, with their skin resembling gooseflesh. (This observation accounts for the term "cold turkey" to describe the withdrawal syndrome.) Efforts to avoid withdrawal symptoms help maintain the addiction. Even more important is compulsive drug seeking behavior, in which addicts return to a drug years after they cease experiencing withdrawal symptoms.

A number of themes recur throughout the history of the major addictive substances. Most began as folk medicines harvested from naturally occurring plants. Most were first regarded by Western physicians as having immediate therapeutic value. And most eventually had their active chemical ingredient isolated, leading to more efficient medical therapy and to a more addictive pure chemical.

Extracts of the poppy plant have been used for the past three thousand years primarily for medicinal purposes, to relieve pain, to calm anxiety, to promote sleep and to ease diarrhea. In 1806, the German chemist Frederick Serturner isolated morphine as the plant's active ingredient. The availability of pure morphine was followed some years later by introduction of the hypodermic syringe. Injectable morphine was a great boon to the rapid treatment of even the severest pain, and was first widely employed in the U.S. Civil War. Of course, injection also sped the psychoactive effects and facilitated addiction, so that by the late nineteenth century, opiate addiction had become known as "the soldiers disease." In a search for less addictive agents, in 1898 the Bayer Company, which would soon introduce aspirin, began marketing a "non-addicting opiate" it called heroin.

Though we generally think of alcohol and tobacco as recreational agents, they, too, were originally employed as medicine. In the late sixteenth century, Jean Nicot, the French Ambassador to Portugal, relayed to the French Royal Court the remarkable therapeutic effects of tobacco. His success as a courtier was

◄ ▲

assured when he used the drug to cure the migraine headaches of Catherine De Médicis, wife and Queen of Henry II of France. When the major active ingredient in tobacco was isolated in 1829, it was dubbed nicotine in honor of Ambassador Nicot's contributions.

The coca plant has been used since antiquity by the Peruvian Indians, providing them with energy to carry heavy bundles through the high altitudes of the Andes. Cocaine was isolated as the active ingredient of the coca plant in the late nineteenth century, and characterized clinically by Sigmund Freud some ten years before he first began practicing psychoanalysis. Freud also used cocaine to relieve depression and enhance mental function. His works on cocaine led to widespread medical use and soon to an appreciation of the drug's extraordinary addictive properties.

In more recent years, many other addictive agents have been synthesized by organic chemists—including barbiturates, Valium-like tranquilizers and amphetamines.

How does one discover the molecular mechanism whereby a drug exerts its effects in humans? This task is simplified if a drug acts through a single specific molecular target designated as a receptor. Receptors are proteins found on the surface of cells to which a drug binds in a very specific fashion, much as a key fits into a lock. Learning the major properties of a receptor clarifies how the drug acts. In the case of opiate receptors which occur on nerve cells or neurons in the brain, mapping their heaviest populations provides insights into the drug's major pharmacological actions. For instance, areas in the brain long known to be involved in processing information about pain are rich in opiate receptors, explaining the analgesic effects of opiates. Euphoric reactions stem from the abundant opiate receptors in parts of the limbic system, which regulates emotional behavior. And opiate receptors are highly concentrated in sites in the midbrain, which regulates the diameter of the pupils of the eye. This concentration explains why opiate addicts invariably have tightly constricted pupils.

The dramatic properties of opiate receptors very much resemble the properties of receptors for neurotransmitters. Neurotransmitters are chemicals released by neurons to act upon adjacent cells, accounting for information processing throughout the brain. Since humans are unable to synthesize their own morphine, some naturally occurring opiatelike substance must function as a neurotransmitter. John Hughes and Hans Kosterlitz in Aberdeen, Scotland were first to identify the brain's own "morphine" as two small peptides they designated enkephalins. Enkephalins are major neurotransmitters that mediate pain perception and emotional behavior. Since enkephalins are endogenous—that is, naturally synthesized—morphinelike substances, many scientists refer to enkephalins and related compounds as endorphins.

Cocaine is a premier example of the stimulant class of drugs and facilitates action of those neurotransmitters that cause a positive emotional state. As the neurotransmitter whose effects are most closely linked to cocaine, dopamine is released by two classes of neurons serving very different functions.

One group regulates movement of the limbs. When these neurons degenerate, the result is Parkinson's disease, which disrupts coordination and often causes a characteristic tremor. By facilitating how dopamine acts in these neurons, cocaine would actually have some therapeutic benefit to Parkinson's victims. Because it is so strongly addictive, however, other, safer therapeutic agents are used in its place.

The other group of neurons that release dopamine regulate emotional behavior. By facilitating dopamine's effects on the emotions, cocaine causes intense euphoria. Individuals who have used many psychoactive drugs invariably regard the cocaine "high" as the most intensely pleasurable state they have ever experienced.

But how does cocaine actually help put dopamine to work? After it is released by a neuron, the dopamine neurotransmitter stimulates the receptor on adjacent cells. Its effect on the receptor is then brought to a halt by a molecular transporter that "pumps" the dopamine back into the nerve that released it. Cocaine blocks this transporter, however, so that more dopamine is available to stimulate cell receptors. As Dr. Michael Kuhar of the National Institute on Drug Abuse has shown, the power of a cocaine derivative is closely related to its ability to block the dopamine transporter.

With insight into the molecular behavior of addictive drugs continuing to accelerate, we hope to develop new agents that can control or arrest it. For example, drugs that block the effects of opiates have already been synthesized, and similar work is under way to deal with the epidemic of cocaine addiction. What's more, in principle it should be possible to develop drugs that block the molecular action of nicotine. If these efforts prove successful, the next 150 years will witness a decline in the addiction that has been a hallmark of the past century and a half.

George Bernard Shaw

was writing plays

and essays in his

nineties. Grandma

Moses began painting

when she was in

her seventies.

NOT GETTING ANY YOUNGER

Declining mortality and morbidity rates over the past century and the consequent increase in the number and proportion of the older population have posed major challenges and sparked intense interest in the scientific study of aging. Although life expectancy at birth, and more recently at age 65, has increased, human

Carl Eisdorfer

longevity is fundamentally unchanged. We are now more likely to live closer to the full measure of our genetic endowment— variously estimated to be from eight to twelve decades, but genotypic life span has not been altered. While changes in sanitation, nutrition, immunization, and antibiotics have played a role, fundamental knowledge of aging has thus far added little to either the length or the quality of our lives.

A variety of theories have been advanced to account for aging and longevity, ranging from magical to moral, endocrinological to molecular, and from purposeful to accidental. Aging is seen in cell loss and reportedly takes one of two forms—cell death (necrosis) secondary to trauma and apoptosis, a response to time-related withdrawal of support factors or some developmental imperative, including the possibility of genetically programmed cell death.

◄

GEORGE BERNARD SHAW

A significant milestone in biologic gerontology (the study of aging as a process) was ushered in with the work of the American scientist Leonard Hayflick on the finite reproductive capacity of cells. Hayflick demonstrated that uncontained cell replication reported in earlier studies was an artifact of the contamination of the cultures under observation. He went on to show that the number of cell replications was in fact limited. It is now agreed that normal cells have a finite reproductive capacity, with cells from older organisms having diminished reproductive life. Since cell replication is normally limited, there is the possibility of a specific gene which controls the number of cell divisions. This is the hypothesized "aging gene," whose discovery has become a goal in a number of laboratories, but which remains still elusive. Perhaps only the cancer cell can be thought of as immortal in its capacity for persistent replication. Understanding factors mediating cancer-cell reproduction can be expected to contribute to our knowledge of aging.

It has been proposed that since the laws of natural selection would be expected to affect a species' characteristics only through reproduction, then traits arising in the post-reproductive period would have no genetic significance. Since longevity expresses itself following the end of reproductive capacity, there is presumably no mechanism for maintaining the trait across generations. If longevity was relevant to species survival, there might be a distinct advantage to groups living at marginal subsistence to have a brief post-reproductive life, thus reducing competition for food or space. Alternatively, there are reports that among some higher-order primates, older females stand guard at the perimeter of their family's territory ready to warn of approaching predators, thus protecting the group.

What's more, some genetically induced processes which are useful in early life may have negative effects over time. The hormone estrogen, for example, which plays a role in reproduction, increases the risk for cancer in later life. This effect is only one instance of antagonistic pleiotropism—the capacity of a gene to produce different phenotypes, some of which may eventually have an adverse effect. We still do not know whether such post-reproductive events are determined or an incidental consequence of an unrelated process.

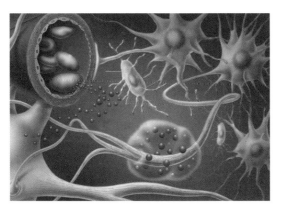

►

Over time there are a number of side effects of metabolism, including the accumulation of destructive material in tissue. Such material, in the form of free radicals (those with unpaired electrons), can oxidize and thus alter a range of molecules and somatic tissues. It has also been shown that the long-term effect of glucose metabolism on collagen creates destructive cross linkages which are dysfunctional.

The American biologist Michael Rose has bred drosophila melanogaster (fruit flies) to a life-span twice their normal length. Rose demonstrated that a gene promoting an anti-oxidant (superoxide dismitase) may enhance longevity in the fruit fly, presumably by altering the effect of a free-radical superoxide.

Environmental stimulation early in an animal's life may also play a significant role in later life, through modifying the production of steroids which are neurotropic and neurotoxic over time. The shorter life expectancies of men compared to women have been attributed to environmental as well as physiological alterations over time. Reports of sex differences in autonomic reactivity and differing psychological performance may also relate to the adaptive capacity. Efforts to lengthen life through diet, smoking cessation, and other health-promoting routines can be effective. Physical exercise is one among a number of extrinsic factors which may indirectly effect longevity. Physical exercise improves aerobic power and muscle, bone structure and cognitive capacity. For humans, perceiving oneself as being independent and in control, as opposed to helpless, has a positive effect. Close social ties enhance both longevity and, apparently, the quality of life.

There are conflicting predictions concerning the potential for morbidity in a lengthened life. It has been thought that, with increasing life-expectancy, diseases will be prevented with a net compression of morbidity, or, alternatively, that as death is postponed by preventing one illness, other debilitating conditions will emerge. Evidence for both points of view exist; we live longer, have healthier productive lives, and have more infirmities, thus increasing the societal caregiving burden. The achievements of contemporary biomedicine in keeping alive more individuals with genetic disorders also adds to this ongoing challenge, but we have little data yet on how things will work out over time.

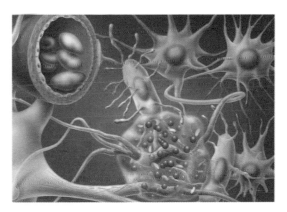

◀

... WHICH WITH
OTHER PROTEINS
GRADUALLY CAUSE
NEARBY NEURONS
TO DEGENERATE.

Studies of learning, work, and performance, including the demonstration of computer skills, show that the proverb regarding old dogs and new tricks is very much in error. Much depends on which skills we teach, how we teach them and the time limits imposed. Such cognitive losses as are normally observed in later life are at a level far lower than conjectured, and dementia—the loss of cognitive capacity such as to impair function—is not a reflection of normal aging per se but is secondary to disease processes affecting the brain. Research into Alzheimer's disease shows it to be probably more than one disorder.

Experience has a facilitating effect on performance which may more than compensate for fundamental age-changes in ability. This balance between reduced ability and enhanced experience has different functional consequences, depending upon the task to be performed. Data demonstrate that the aged show considerable variance in psychological as well as physiologic characteristics.

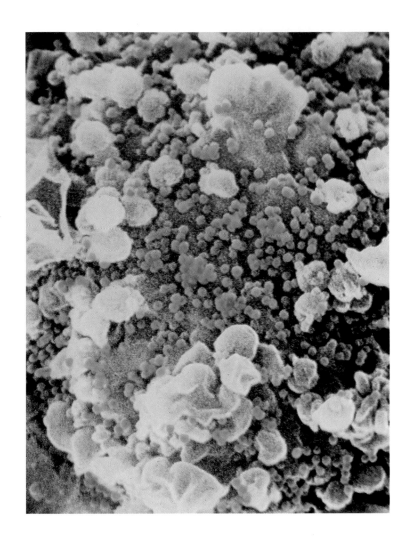

THE PANDEMIC SPREADS

The origins of AIDS and its causative agent, the human immunodeficiency virus (HIV), remain obscure. Yet based on current evidence, it appears that the worldwide spread of HIV began in the mid-1970s. By 1980, an estimated 100,000 people

Jonathan M. Mann

worldwide had become infected; during the 1990s, this number increased a hundredfold, to about 10 million. Truly a world-wide epidemic (properly termed a pandemic), AIDS cases have now been officially reported from 187 countries, and the number of people estimated to have been infected with HIV

exceeds 23 million, including about two million children.

The HIV pandemic is rather new, and as such, it remains a dynamic and volatile phenomenon. In addition to its continued spread through previously affected countries (with an estimated 40-80,000 additional HIV infections in 1994 in the U.S. alone), HIV is moving on to communities and countries that were once spared. Today, over 90 percent of HIV-infected people live in the developing world. The most explosive recent spread has been in Southeast Asia, where at least two million Indians, 500,000

Thais and 300,000 Burmese have become HIV infected in just the past few years. Indonesia, Bangladesh and Pakistan may well be next, and global estimates for the year 2000 range between 40 and 110 million HIV-infected people.

◀

A VIRUS KNOWS NO SOCIETAL ISSUES OR BIASES. HIV WILL INFECT ANY AND ALL T LYMPHOCYTES IT ENCOUNTERS.

With time, the epidemic has become increasingly complex. In the U.S., an epidemic that had predominantly affected gay men in urban areas of a few states has become an epidemic that increasingly infects heterosexuals, injecting drug users, ethnic minorities, the urban poor and rural areas. Most large cities now have multiple, simultaneously occurring HIV epidemics in different population groups.

The impact of HIV/AIDS has been profound and broad, affecting virtually all sectors of society, from agriculture to education, health care and social services. Medical-care costs are high (estimated at a global total of $5 billion for 1992) and indirect costs from the loss of young and middle-aged adults have been enormous.

The major impact of the HIV/AIDS pandemic is yet to come. Since AIDS does not develop, on average, until about 10 years after initial infection with HIV, the number of people with AIDS increases dramatically over time. For example, while an estimated five million people have developed clinical AIDS through early 1994, this number will increase in just two years by about 60 percent, as nearly three million additional cases occur. Thailand, which experienced at most several hundred AIDS cases over the last five years, must prepare for at least 100,000 AIDS cases during the next five years. By all criteria—the number of people infected with HIV and with AIDS, the number of children orphaned by AIDS, the growing stresses on our health and social welfare systems—the decade to come will be much more difficult than the decade past. The pandemic is expanding and spreading, becoming ever more complex and altering the social landscape around the world. But this is only part of the story. Our response to the pandemic will determine its future.

The pace of scientific discovery and understanding has been remarkable. The recognition of the new disease—AIDS—in 1981 was itself extraordinary. Discovery of

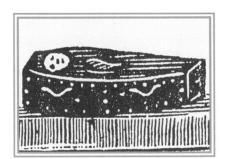

▶

IN THE U.S. ALONE, AIDS HAS CLAIMED MORE THAN 200,000 LIVES.

the causative virus, HIV, and research into mechanisms of HIV infection and AIDS has led to enormous achievement (and, admittedly, frustration) as well as to the first treatments for both AIDS-related opportunistic infections and cancers. Still, the search continues for effective antiviral agents and a protective vaccine.

The lesson from global experience is that prevention works. Effective local prevention programs require a three-pronged approach—information and education; health and social services, such as confidential counseling and testing, condoms, drug treatment programs and sex education; and, not least, a supportive social environment. This combined approach has proven effective in many countries

Gay and Straight

In the early 1980s, AIDS was often referred to by the media as a "gay plague." This label contributed to delays in recognizing and responding to AIDS in many countries. Yet this unfortunate fixation on AIDS as a homosexual disease resulted from the historical accident that AIDS was first recognized among gay men in the United States. For had AIDS been discovered *initially in Africa, where heterosexual spread dominates (and where AIDS was certainly present in the late 1970s), it would have been known as a sexually transmitted disease which also affects homosexuals. Overall, about 70 percent of HIV infections worldwide are heterosexually transmitted, about five times the number of homosexually transmitted infections.*

and for many different groups—everyone from adolescents, adults, and homosexuals to heterosexuals, injecting drug users and sex workers.

The social and political commitment to fighting AIDS is, unfortunately, declining worldwide. What is being done today against AIDS is vitally important, but it is not sufficient to bring the pandemic under control. Paradoxically, since 1990 there has been a dangerous, rapidly widening gap between the pace of the pandemic and the strength and robustness of the global response.

A decade of experience has shown that vulnerability to HIV has a societal dimension. The pandemic is gravitating toward those who are marginalized, discriminated against and stigmatized within each society. In order to uproot the pandemic, pre-existing, underlying societal issues such as women's inequality must be addressed.

Therefore, in a totally unexpected manner, the HIV/AIDS pandemic is bringing about dramatic changes in how we understand health and its relationship to society. An awareness is dawning that the effort to promote and protect health cannot be separated from promoting and protecting human rights and dignity.

◄

WEARERS OF RED RIBBONS DISPLAY THEIR COMPASSION FOR AIDS VICTIMS AND THEIR FAMILIES.

AGE OF GRAND DISCOVERY

1850

GERMAN PHYSICIST RUDOLF CLAUSIUS publishes an article presenting the second law of thermodynamics. He finds that in every energy conversion some energy becomes heat and that heat energy can never be converted completely to any other form of energy. The energy supply of the universe, therefore, is constantly being degraded. In trillions of years, Clausius concludes, no useful energy will remain.

52

BRITISH PHYSICISTS JAMES JOULE AND WILLIAM THOMSON —later to become Lord Kelvin—establish that an expanding gas becomes cooler. This becomes known as the Joule-Thomson effect.

LORD KELVIN

U.S. ENGINEER ELISHA OTIS opens the sky to skyscrapers by developing the first passenger elevator with a safety guard. The guard prevents the elevator from falling even when the cable is cut, and it gives developers the confidence to place elevators in their buildings. Otis installs the first safe and workable elevator in New York's Crystal Palace two years later.

53

BRITISH ENGINEER GEORGE CAYLEY designs the first glider capable of sustained flight in rising updrafts. Too old to test the contraption himself, however, he orders his reluctant coachman to do so. The coachman flies 500 feet and escapes uninjured.

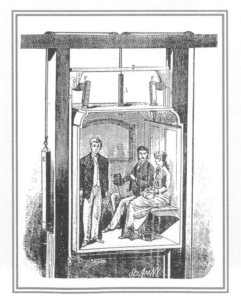

◀ AN EARLY PASSENGER ELEVATOR

54

WITH A CHOLERA EPIDEMIC RAGING IN LONDON, British physician John Snow finds hundreds of cases within a few blocks of a water pump that is near a sewer pipe. When he removes the pump handle, the incidence of cholera decreases, giving emphasis to improving hygiene as a means to disease prevention.

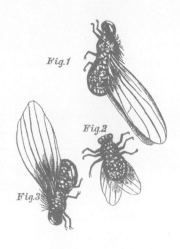

Fig.1

Fig.2

Fig.3

▲

CHOLERA WAS ONCE ATTRIBUTED TO THE INHALATION OF SMALL POISONOUS FLIES.

56

FRENCH CHEMIST LOUIS PASTEUR investigates why wine turns sour as it ages. He finds yeast cells to be the culprit and suggests that wine makers gently heat their product to kill the yeast before corking it. Vintners discover that pasteurization works, as do dairy farmers, who eventually apply the process to milk.

THE WORLD OF FASHION CHANGES forever when British chemistry student William Perkin fails in an attempt to synthesize the antimalarial drug quinine.

The solution Perkin creates has a purplish tint that is later called mauve. Enamored with the color, Perkin finds that it dyes clothes. He launches the field of synthetic dyes, a field that catches on so quickly that the next few years in England are known as the Mauve Age.

58

THE FIRST transatlantic telegraphic cable is laid.

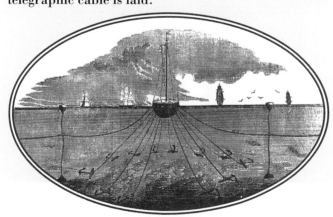

59

BRITISH NATURALIST CHARLES DARWIN reluctantly publishes *The Origin of Species*, which sets forth his principle of natural selection and its influence on evolution. Though he has spent 20 years researching the theory, he knows that his proposal will stir controversy and wants to spend still more time gathering evidence. By 1858, however, British biologist Alfred Wallace has sent Darwin an 11-page letter outlining the same theory. Darwin concludes he has no choice but to immediately propose joint publication. Many scientists quickly embrace the theory, but the public, wary of its atheistic overtones, is less enthusiastic.

FRENCH INVENTOR JEAN-JOSEPH LENOIR opens the door for the development of the automobile and the airplane when he builds the internal-combustion engine— a device far less bulky than a steam engine.

▲

THE ORIGIN OF SPECIES'S FIRST EDITION OF 1,250 COPIES SOLD OUT ON THE FIRST DAY OF PUBLICATION.

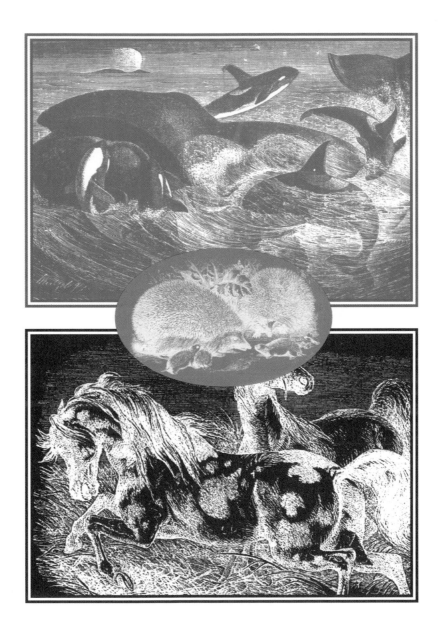

THE LIVELIEST ASPECT OF ALL THAT LIVES

The study of animal behavior is among the oldest of human endeavors. As hunter-gatherers, our ancestors needed intimate knowledge of both flora and fauna, including the habits of their prey. A more practical kind of knowledge became necessary when our ancestors took up agriculture, and began to keep animals for food and work. The hunter's orientation to natural behavior, and the farmer's utilitarian perspective, are still recognizable today in the two main scientific approaches to animal behavior—that of the zoologist and the psychologist.

Zoologist Konrad Lorenz, the Austrian pioneer of ethology, called behavior "the liveliest aspect of all that lives." Observing animals under natural or naturalistic conditions, ethologists are interested in their characteristic behavior, such as how they defend territories, court the opposite sex, catch prey, escape predators, care for their young, communicate with each other, and so on.

Frans B.M. de Waal Given the millions of species on earth, the variability and grounds for comparison are inexhaustible. However, ethologists have discovered many principles that show up in disparate animal families. For example, once the discovery of echolocation solved the mystery of how bats orient themselves in the dark, similar sonar capacities were found in dolphins and even in some cave-dwelling birds. These animals emit ultrasonic sounds that, once reflected back, allow them to detect a variety of objects, including prey.

Ever since the days of Charles Darwin, biologists have been particularly interested in behavior as an adaptation to the environment. Zoologist Nikolaas Tinbergen, the Dutch cofounder of ethology, conducted elegant experiments to test evolutionary hypotheses. For example, he wondered why many birds remove egg shells from the nest after their chicks have hatched. Do they do so because chicks hurt themselves on the shells' sharp edges? Or is it because eggs are camouflage-colored on the outside but not on the inside which is exposed once the chicks have hatched? Tinbergen confirmed this second function: predators, such as crows, are more likely to find unhatched eggs if there are empty shells placed next to them. The nesting birds themselves do not need to learn this costly lesson: egg-shell removal is an automatic response favored by natural selection, because birds that do it have more surviving offspring than birds that do not.

The attention of ethologists is not limited to such "instinctive" behavior—that is, behavior that is genetically determined and relatively inflexible. Indeed, one of the early triumphs of this science was the discovery of a learning process known as imprinting. Hatchlings of certain birds, such as geese and chickens, attach themselves to the first moving object they lay their eyes on. Because normally this is their mother, they develop a preference for their own species. In experiments, however, the imprinted object can also be a toy truck.

The tendency to acquire crucial and indelible knowledge in early life is widespread in the animal kingdom. Critical learning periods, as they are known, suggest a predisposition to absorb specific information at a specific age, as if the organism's neural networks are waiting for it. In our own species, language acquisition is thought to be an example of such a learning disposition: after a certain age, it becomes virtually impossible to attain native levels of language proficiency.

Behaviorists differ from ethologists in that they are not interested in animals for their own sake, but study them to find general laws of behavior and, ultimately, to help us understand ourselves. Under the influence of American comparative psychologists, such as E.L. Thorndike and B.F. Skinner, they use white rats, pigeons, and other domesticated species as laboratory "models" for human behavior. Learning through trial and error is the main object of study.

Crucial for the development of this science was the discovery, early in this century, of the conditioned reflex by the Russian physiologist Ivan Pavlov. Pavlov demonstrated that if a dog hears a bell each time it is about to get food, after a while, it will salivate at the sound of the bell alone. The dog is "conditioned" to the sound by associating it with food.

A most promising development in recent decades is sociobiology. Biologists have always been puzzled by the altruistic behavior of animals, such as ant soldiers that give their lives in defense of their colony or mongooses that expose themselves on top of a termite mound to warn group mates of approaching danger. One would think that individuals that care more about their own survival than about that of others would leave more offspring, thus spreading their genes more widely in the population. How, then, could altruistic behavior have evolved?

Assuming that natural selection favors assistance that serves an animal's genetic relatives or is repaid through return-favors, sociobiologists have been quite successful in explaining the cooperative nature of animal societies. The highest degree of cooperation is indeed found among close kin, and there is increasing evidence that some animals, such as monkeys and bats, reciprocate the services they receive from others. Although extending these theories to human behavior has raised great controversy, sociobiology has developed a growing following among psychologists and anthropologists.

The last major field to emerge in the study of animal behavior is known as cognitive ethology. This field addresses such questions as: Do animals realize the effects of their own behavior? Do they understand relationships in their group in the same way people do—such as who is married to whom or who has higher status?

Placing less emphasis on the learning of behavior—as behaviorism does—than on information processing capacities and the animal's understanding of its own environment, cognitive ethology has brought new insights to animal communication (monkey alarm calls, for example, can indicate not just the presence of danger but also distinguish between aerial and ground predators), in the "politics" of social life (rather than through brute strength, chimpanzees attain high rank through alliances with others—hence their need to rally support before challenging the existing order), in self-awareness (only apes and humans appear to recognize their mirror-image as themselves), in deception (one animal may trick another, for example, by presenting a relaxed appearance that lets it attack when the other has come too close), and even peacemaking after fights.

Kissing and Making Up

Golden monkeys do it with mutual hand-holding, chimpanzees with a kiss and embrace, bonobos with sex, and tonkeana macaques with clasping and lipsmacking. Each species follows its own peacemaking protocol. Many have evolved gestures, facial expressions, and calls specifically for this purpose, while other species, lacking such refinements, display reconciliations that look much the same as any other contact.

How can we know, then, when a species is reconciling? Because the criterion is not a particular behavior, but a particular sequence of events defined as a reunion between former opponents not long after a fight. Thus, a few accidental reunions are not enough: reconciliation is believed to occur only if there is a systematic increase in friendly behavior following aggressive exchanges.

Quite a few studies have detected reconciliation behavior and concluded that former opponents are selectively attracted—that is, they tend to come together more often than usual, and more often with each other than with individuals who were not involved in the fight. These reconciliations serve to "undo" the damaging effects of aggression and seem widespread in the primate order.

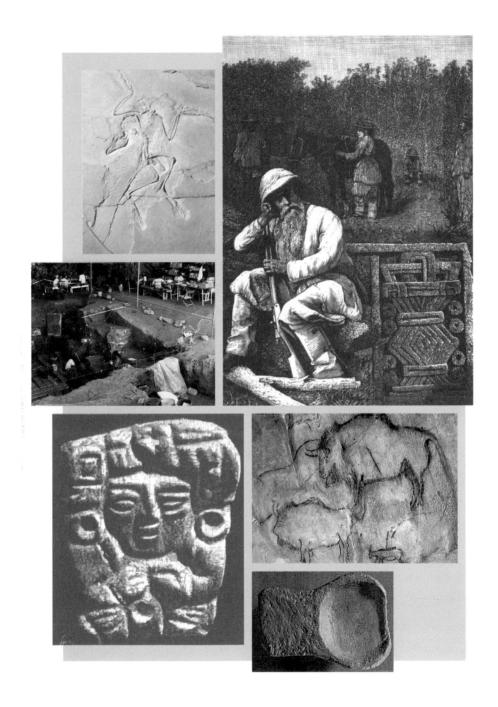

DIGGING INTO THE PAST

The last 150 years has witnessed the emergence of archaeology as a professional discipline throughout the world. It has been a time of extraordinary growth in archaeological knowledge of the past and in the development of sophisticated techniques and methods to retrieve and interpret the past.

By 1845, Christian Jürgensen Thomsen, the first director of the Danish National Museum, had long since established the three-age system of classifying

Jeremy A. Sabloff

archaeological artifacts into those of the stone age, bronze age and iron age; John L. Stephens (US) and Frederick Catherwood (UK) had published their careful descriptions and drawings of ancient Maya ruins; Jacques Boucher de Perthes of France had begun to publish reports of his discoveries of hand axes and the

remains of extinct animals in early geological contexts in the Somme Valley; and Ephraim G. Squier and Edwin H. Davis (both US) were shortly to publish the results of their pathbreaking surveys of archaeological sites in the greater Mississippi Valley. Soon after, archaeological research around the globe rapidly expanded, as a number of national and local museums were founded and archaeology became a full-time pursuit. Archaeological sites and artifacts were uncovered, described, and classified, as expanding knowledge about the human past began to be ordered in chronological frameworks.

In the wake of the Darwinian revolution of the mid-nineteenth century, scholars began to classify both modern and ancient cultures into evolutionary sequences. Unfortunately, many of these putative evolutionary schemas, for example, "savagery," "barbarism" and "civilization," were as racist as they were unscientific, and thus began falling out of favor by the close of the nineteenth century.

By the early years of this century, archaeology was thriving. Geographical and chronological gaps in our archaeological knowledge began to be filled in through new excavations and surveys. Much attention was paid to artifact analyses, particularly of pottery (the ubiquitous potsherd) and stone tools. These trends accelerated after World War I. Through the efforts of outstanding scholars such as Australian archaeologist V. Gordon Childe and Alfred Kroeber (US), new attention was drawn to the overall cultures of ancient peoples, their development, and the movement of peoples and diffusion of cultural traits over the landscape. The influence of the environment as an active cause of cultural change also began to receive more attention.

◀

THE PAGES OF HISTORY AND PRE-HISTORY HAVE UNFOLDED FOR ARCHAEOLOGISTS OF ALL ERAS. THEIR FINDINGS INCLUDE ITEMS SUCH AS THE FOSSIL OF AN EARLY PREDECESSOR OF MODERN BIRDS (*TOP LEFT*) AND CAVE PAINTINGS (*MIDDLE RIGHT*) THAT WERE CREATED BY THE LIGHT OF A PALEOLITHIC LAMP (*BOTTOM RIGHT*).

By the middle of this century, the invention of radiocarbon carbon-14 dating by American chemist Willard Libby made chronological ordering of the past far easier and more accurate. Use of this dramatic new system spread rapidly throughout the world as archaeologists suddenly found themselves able to assign absolute dates to previously undatable prehistoric materials. In this half of the century, still more scientific tools have been added to the archaeological arsenal, and archaeological research both in the field and in the laboratory has increasingly become a collaborative venture between archaeologists and a wide variety of scientists—among them geologists, chemists, physicists and botanists. Collaborations with other scholars, such as linguists, historians, and architects, have also become common.

By the beginning of the 1960s, archaeologists were increasingly able to decipher some of the major transitions in human history and to reconstruct them in greater depth than ever before. For example, scholars attained a stronger grasp of the evolution of the genus Homo from its earliest origins in Africa; of the lifeways of the mobile hunter-gatherers of the Late Pleistocene and early Post-Pleistocene period; of the rise of plant and animal domestication and the beginnings of settled village life; and the growth of complex societies and the rise of cities in the ancient world. Indeed, researchers unearthed more information about the great early civilizations of the Near East, Africa, the Indus, China, Mesoamerica, and the Andes than ever before, with their work encompassing not merely the ruined temples and palaces of the elite, but the remains of peasant houses and activities as well.

In the second half of this century, the main intellectual change in archaeology, principally within the specialization of anthropological archaeology, has been the rise of what is called "processual archaeology" or the "new archaeology." The main tenet of this new school of thought is for archaeologists to strive to understand the nature of cultural change over long periods of time, and to develop tools to enhance their ability to gain such an understanding. In effect, most archaeologists today have shifted their focus from questions of "what," "where," and "when" to questions of "how" and "why." Placing artifacts—the material remains of the past—in time and space has thus become simply a means of solving the puzzle of how and why cultures adapt and change through time. Although there have been some criticisms of this new approach, in particular its overemphasis on the environment and technology at the expense of religion and ideology, and of archaeology in general, especially its frequent unwillingness until recently to collaborate with native peoples around the world, most archaeologists are optimistic, as this century draws to a close, that the field is poised to productively use its vast database of knowledge about past cultures to answer significant questions about why human society has developed along the paths that it has.

Many archaeologists in the field are confident that answers to such questions as "What was the adaptive value of the rise of large urban places?" or "How were large populations able to sustain themselves over long stretches of time in difficult environments like tropical forests?" are not just of antiquarian interest, but relevant to crucial problems that confront us in the modern world.

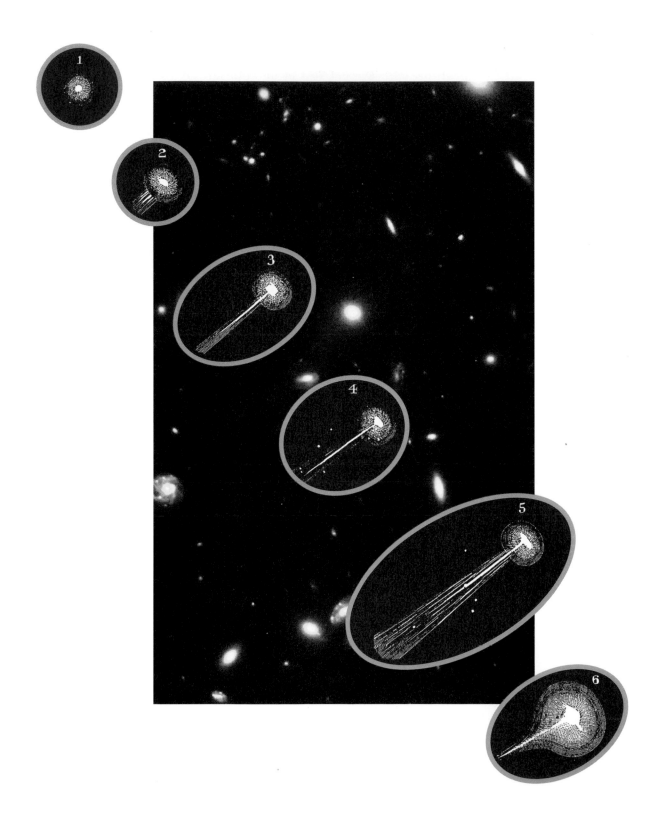

OBSERVING THE UNIVERSE

Astronomers in 1845 spoke of the "fixed" stars in contrast to the wandering planets and had a good understanding of the positions, distances, and brightnesses of both. They were, however, essentially ignorant of the chemical make-up of the sun and stars, of the energy sources that keep them shining and of the structure of the universe (or even its existence) outside of our own Milky Way galaxy. No one had ever photographed a star or the informative rainbow (spectrum) produced when the light of a star passes through a prism or a grating, and no one had ever used a telescope sensitive to anything except ordinary visible light.

Virginia Trimble

The most important advances in observational astronomy in the intervening 150 years have been the development of spectroscopy, beginning in the 1880s, with its potential for measuring chemical compositions, temperatures, and motions of astronomical objects; the use of photographic plates (from about 1890) and of electronic detectors (starting slowly in the 1930s and taking off in the 1970s) to record images and spectra, permitting study of much fainter objects and accurate comparisons across space and time; and the opening of new windows in addition to visible light, at radio, x-ray, infrared, ultraviolet, and gamma-ray wavelengths, between about 1945 and 1985, with the development of detectors for neutrinos and gravitational radiation just getting under way now.

These advances in astronomical technology have been matched by the addition of powerful theoretical tools, beginning with Maxwell's equations for the behavior of electric and magnetic fields and light in 1878 and continuing with quantum mechanics (1900-1925) and quantitative descriptions of the structure and behavior of atoms, their nuclei, and interactions among them (1925 to the present). Einstein's special (1906) and general (1916) theories of relativity have also proven important for understanding astronomical processes under extreme conditions, when objects are very big, very dense or very hot. Postwar development of the physics of elementary particles (protons, neutrons, their constituents, and electrons) is just beginning to merge with observational cosmology to show how galaxies and the largest-scale structures in the universe must have formed 10 or 20 billion years ago.

As a result of the synergism between observation and theory, we can now say with some confidence that we know why stars shine, have a good picture of the large-scale structure and evolution of the universe (cosmology), and recognize

◄

THE GALAXY AS SEEN THROUGH THE HUBBLE TELESCOPE (*BACKGROUND*). ILLUSTRATIONS OF THE OBSERVATION OF PROSPER HENRY'S COMET AS SHOWN IN *SCIENTIFIC AMERICAN* IN 1894 (*INSETS*).

that the universe and objects in it are far more changeable and even violent than our ancestors saw or could have seen.

The answer to the stellar puzzle is that typical stars like the sun are made mostly of the lightest two elements, hydrogen and helium (with only about two percent made up of the more familiar substances like carbon, oxygen, iron and gold) and that they draw their energy from nuclear reactions in their cores, especially the fusion of hydrogen to helium. The correct idea was initially advocated before 1926 by Sir Arthur Eddington, based on laboratory measurements (by F.W. Aston) of the masses of hydrogen and helium atoms, in combination with Einstein's equation, $E = mc^2$. Hydrogen fusing to helium could power stars for billions of years, if there were enough of the elements. Cecilia Payne Gaposchkin recognized (1925) and Henry Norris Russel confirmed (1929) that stars really are made mostly of hydrogen and helium. The correct nuclear reactions were written down in 1939 by Hans Bethe, who won the 1968 Nobel Prize in physics. Since then, the structure and evolution of stars has become the best-understood part of astrophysics.

The universe has turned out to be enormously larger than our own galaxy (perhaps infinitely large), old and expanding rather than static. Edwin Hubble (from 1924 onward) used both imaging and spectroscopy of the so-called spiral nebulae to show, first, that they were not mere clouds of gas in the Milky Way, but rather galaxies in their own right, each containing billions of stars like the sun, and, second, that they and we are all moving apart from each other, and the most distant galaxies are moving the fastest. This relationship between distance and speed—Hubble's law—in combination with Einstein's general relativity requires that the entire four-dimensional space-time we live in (the universe) be expanding uniformly.

The age of the universe is roughly the time scale of that expansion, and the value of 10-20 billion years is confirmed by the ages of the oldest stars and by the decay rates of uranium, thorium, and other radioactive elements. Measuring these requires a combination of observations of stars, data from terrestrial laboratories, and theoretical nuclear physics.

The long-term future of the universe remains uncertain. It might either eventually fall back onto itself or expand forever, depending on whether the average density is larger or smaller than a critical number. The measured density is so close to the dividing line that we cannot be sure whether the life expectancy of the universe is finite (though hundreds of billions of years in any case) or infinite.

Finally, each new wavelength window opened has shown us new kinds of astronomical objects and required new physical concepts for their explanation. The first radio astronomers found much brighter sources than anyone had expected, coming from the centers of violently active galaxies, where black holes probably lurk, and from the remains of stellar supernova explosions. The residual cores are rapidly rotating, highly magnetized neutron stars (pulsars). The ejecta include particles moving nearly at the speed of light, probably related to the relativistic particles or cosmic rays found everywhere in the galaxy, including near the earth (where they cause aurorae, mutations, and other effects).

The first x-ray astronomers found additional neutron stars and the first stellar mass black holes, both bright sources when gas from companion stars falls down onto them. Additional, more violent active galaxies also appear in x-ray catalogues, some changing their brightnesses in only hours. Infrared astronomy has probed the short-lived phases of star formation and revealed galaxies temporarily brightened by collisions with their neighbors. Opening the gamma-ray window has required not only satellites to rise above the earth's atmosphere but also new kinds of detectors. (Your eyes will record gamma rays, but at the price of permanent damage.) Most mysterious of the new sources are the gamma-ray bursters. These flare in seconds or less, die away completely, can be seen at no other wavelength and cannot at present be convincingly identified with any known sort of object inside our galaxy or out of it.

"Twinkling" Sources

Many astronomical advances illustrate chance favoring the prepared mind (as Pasteur said). In 1967, Anthony Hewish and his colleagues at Cambridge University built a new sort of radio telescope, designed primarily to look for compact sources in distant galaxies, conspicuous because irregularities in the gas between us and them can make them seem to twinkle like visible stars. Data was recorded on punched paper tape, and much of the analysis turned over to a graduate student, S. Jocelyn Bell. She identified many of the desired "twinkling" sources, which eventually formed the body of her Ph.D. dissertation. But she also became so skilled at recognizing patterns in her data that she spotted four examples of a new category of source among the hundreds scanned by the telescope, calling them to her colleagues' attention over the 1967-1968 winter holidays. Thus were pulsars discovered. They were soon associated with rotating neutron stars, left behind in supernova explosions, triumphantly confirming ideas suggested in 1934 by Walter Baade and Fritz Zwicky. Hewish is now Sir Anthony, and he shared the 1974 Nobel Prize in physics. Bell is now Prof. S.J. Bell-Burnell of the Open University in England, and many in the astronomical community feel that she did not get her fair share of the credit.

1860

ABRAHAM LINCOLN IS ELECTED sixteenth president of the United States. South Carolina, the most extreme of the slave states, reacts to Lincoln's victory by seceding from the nation.

▲
BOTH NORTH AND SOUTH CLAIMED VICTORY IN THE FIGHT BETWEEN THE *MERRIMAC* AND THE *MONITOR*.

61

ARCHAEOLOGISTS UNEARTH A FOSSIL of Archaeopteryx, a small lizard-like animal estimated today to be 140 million years old. Many other dinosaur fossils have been discovered before and since, but this lizard is unique because of its feathered wings. Half reptile and half bird, it is among the best fossil examples of evolution at work.

IRISH PHYSICIST JOHN TYNDALL proposes that carbon dioxide and water vapor let visible light from the sun reach the earth but block the infrared radiation the earth emits toward space at night. He has discovered what will eventually become known as the greenhouse effect—a phenomenon that will prompt theories of global warming.

62

LOUIS PASTEUR PUBLISHES HIS GERM THEORY of disease, giving credence to the growing perception that contagious diseases are caused by microorganisms. This realization launches modern medicine—prompting Pasteur and others to seek out disease-causing microorganisms and to fight them at their source.

63

THE WORLD'S FIRST SUBWAY system opens to the public in London.

65

CONFEDERATE GENERAL ROBERT E. LEE surrenders to Union general Ulysses S. Grant at Appomatox Court House, ending a Civil War that has cost the United States more than one million casualties. Five days later, actor John Wilkes Booth assassinates President Lincoln.

66

AFTER LOSING A BROTHER to an explosion in the family's nitroglycerin factory, Swedish industrialist Alfred Nobel, seeks to tame nitroglycerin's deadly volatility. He invents dynamite, which combines nitroglycerin with diatomaceous earth and cannot be ignited without a detonating cap.

AFTER SPENDING FIVE YEARS studying the relationship between tall pea plants and dwarf pea plants in his monastery garden, Austrian botanist and Augustinian monk Gregor Mendel publishes his theories of heredity in an obscure scientific journal. Although he has founded the science of genetics, his work will go unnoticed for 33 years.

67

U.S. INVENTOR Christopher Sholes builds the first commercial typewriter.

68

WORKERS BUILDING A ROAD IN FRANCE discover five skeletons in a cave named Cro-Magnon. These Homo sapiens remains are 35,000 years old and become known as Cro-Magnon man.

69

RUSSIAN CHEMIST DMITRI MENDELEEV publishes the first version of the periodic table of elements.

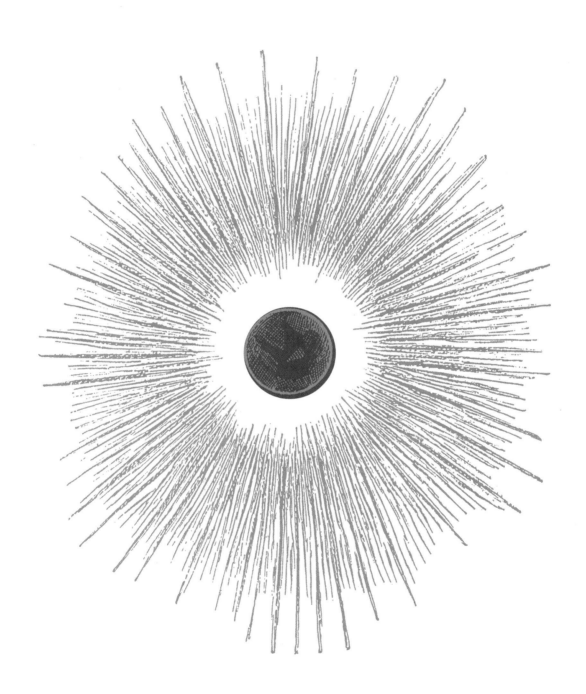

IN THE BEGINNING

One hundred and fifty years ago the first quantitative measurements of the size and age of the universe were being made. In 1838, the German astronomer Friedrich Wilhelm Bessel determined the first distance to a star, finding that 61 Cygni is almost a million times further away than the sun. In 1862, British physicist Lord Kelvin made the first quantitative estimate of the age of the earth, judging it to be about 200 million years old—or almost 30,000 times older than seventeenth century Irish Archbishop James Ussher's biblical deduction, which placed the Creation at 4004 B.C.

Although German philosopher Immanuel Kant had suggested in 1755 that the nebulae, or cloudy patches, seen in the sky were "island universes"—distant star systems like our own galaxy, the Milky Way, rather than objects within the Milky Way itself—his speculation was not confirmed until 1923.

Michael S. Turner In that year, American astronomer Edwin Powell Hubble used the new 100-inch Hooker telescope at the Mt. Wilson Observatory near Pasadena, California to show that the Andromeda nebula is well beyond the Milky Way. Six years later, in 1929, Hubble made an even more momentous discovery, finding that all galaxies are moving away from one another and that the universe is expanding. His work established a beginning for space and time of about 15 billion years ago.

By 1930, the gross features of the universe were known. Galaxies, comprised of hundreds of billions of stars and separated by millions of light years, are the building blocks. They are usually found in small groups containing two to ten galaxies and less frequently in clusters containing as many as several thousand. The largest structures in the cosmos are superclusters, comprised of several clusters of galaxies, and "voids," great regions of space devoid of galaxies. The observable universe is 15 billion light years across and contains some 100 billion galaxies.

In 1964, American physicists Arno Allan Penzias and Robert Woodrow Wilson discovered that the earth is being bathed uniformly by microwave radiation at about three degrees above absolute zero, and thus, that the universe has a temperature. (The National Aeronautics and Space Administration's Cosmic Background Explorer [COBE] satellite recently determined the temperature to great precision, 2.726 ± 0.005 Kelvin.) Since the universe has been cooling as it expands—and today, after 15 billion years of expansion, its temperature is still measurable—its beginning must have been very hot (with temperatures approaching 1032 Kelvin).

At very high temperatures, matter is reduced to its most fundamental constituents, quarks and leptons, and so the universe began as a hot soup of these particles. There is thus a link between the study of inner space, or elementary-particle physics, and the study of outer space, or cosmology.

As the universe evolved from its hot, simple beginning, layer upon layer of structure gradually developed. At 10-5sec the temperature had cooled to 1013 Kelvin and the strong color force bound quarks together into neutrons and protons; at around 300sec the temperature had dropped to a billion degrees Kelvin and the strong nuclear force bound all of the neutrons and some of the protons into nuclei of the lightest elements—hydrogen, helium and lithium. When the universe was 300,000 years old, the temperature had fallen to 3000 Kelvin and the electromagnetic force bound electrons to these light nuclei, creating atoms. Because atoms are relatively transparent to radiation, the heat radiation in the

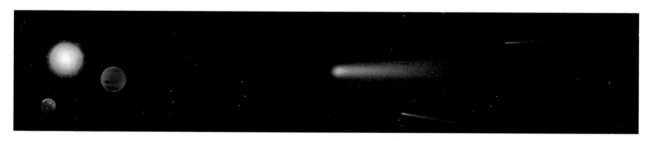

▲

THE DISTANCE
FROM THE EARTH TO
FAR-REACHING COMETS
IS OVER 10 TRILLION
KILOMETERS.

universe streamed freely through them. This radiation is seen today as the microwaves discovered by Penzias and Wilson, and provides a "fossil record" of the early Universe. The final layer of structure, the organization of matter into stars, galaxies, clusters of galaxies, and superclusters, was driven by gravity. Its attractive force amplified very small (about 0.01 percent) inhomogeneities, or "lumps," in the density of the universe into the speckled, nonuniform structure seen today.

The hot big-bang model accounts for the development of the universe from 10-5sec after the bang until the present. It is supported by four observational pillars: (1) the expansion of the universe, (2) the presence of cosmic background radiation (CBR), (3) the agreement between the observed abundances of hydrogen, deuterium, helium and lithium and those predicted by the theory of big-bang nucleosynthesis, and (4) the COBE satellite's discovery of very small variations in the temperature of the CBR (about 30 microKelvin) between different points in the sky, which provides evidence for the primeval lumpiness that seeded all the structure seen today.

Like many successful theories, the big-bang model has allowed us to ask a deeper set of questions, whose answers likely involve events that took place during the very earliest moments. Why, for example, is the universe so smooth (recall the uniformity of the CBR temperature), and why is the universe so flat (Einstein's theory predicts that space-time is curved, yet we see no evidence for large-scale curvature)? What caused the lumpiness needed to seed the structure that exists today? What is the quantity and composition of the ubiquitous dark matter? Most of the matter in the universe is dark, of unknown composition. Why is the universe composed of matter alone rather than equal parts of matter and antimatter?

Some cosmologists believe that we are well on our way to answering these questions and thereby establishing a "grander" big-bang theory. Russian physicist Andrei Sakharov suggested that particle interactions in the early universe led to a slight excess of matter over antimatter. After the antimatter collided with matter and was annihilated, only matter was left. This idea fits nicely into

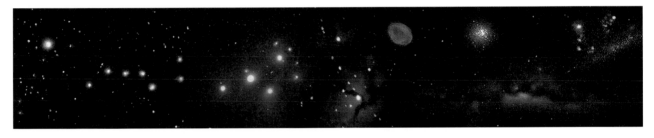

particle-physics theories that unify the strong, weak and electromagnetic forces (grand unified theories). American physicist Alan Guth proposed a small but crucial modification of the big-bang theory, called inflation. At very early times a bizarre form of energy predicted to exist by grand unified theories and called false-vacuum energy drove an extremely rapid period of expansion—in 10 to 32 seconds the universe expanded by a larger factor than it has since the end of inflation. The enormous amount of expansion "inflates" a tiny part of the universe that is smooth and appears flat to a size that encompasses all we see today as well as stretching quantum mechanical fluctuations on subatomic scales to cosmological scales. If this idea is correct, it explains the origin of the primeval lumpiness as well as why the region of the universe we observe is so smooth and flat. As the Russian Andrei Linde has emphasized, inflation has changed our global view of the universe; on extremely large scales, the universe should consist of a patchwork of regions that can never communicate with one another. These regions should have developed very differently, perhaps even with different laws of physics and numbers of space-time dimensions.

The most pressing problem in cosmology today involves dark matter and the details of structure evolution. While we still do not know how much matter there is

▲

THE BIG DIPPER SITS 100 LIGHT-YEARS FROM THE EARTH; 100,000 LIGHT-YEARS SEPARATE THE LARGE MAGELLAN CLOUDS AND OUR PLANET.

Tuning in the Cosmos

The existence of cosmic background radiation (CBR) may be the most important discovery of the century; it certainly involved the most curious sequence of events. In 1948, Russian-born physicist George Gamow, and Ralph Asher Alpher and Robert Herman (US) predicted the existence of background radiation of a temperature of about 5 Kelvin. Though Gamow's pioneering writings about the big bang received much attention, no one took the idea seriously enough to look for actual cosmic background radiation. Its effects, however, had already been discovered by astronomers W. S. Adams (US) and A. McKellar (Canada), who noticed

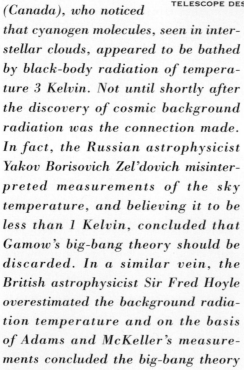

TELESCOPE DESIGN FROM 1874

that cyanogen molecules, seen in interstellar clouds, appeared to be bathed by black-body radiation of temperature 3 Kelvin. Not until shortly after the discovery of cosmic background radiation was the connection made. In fact, the Russian astrophysicist Yakov Borisovich Zel'dovich misinterpreted measurements of the sky temperature, and believing it to be less than 1 Kelvin, concluded that Gamow's big-bang theory should be discarded. In a similar vein, the British astrophysicist Sir Fred Hoyle overestimated the background radiation temperature and on the basis of Adams and McKeller's measurements concluded the big-bang theory was wrong.

In 1964, Arno Penzias and Robert Wilson set out to study radio emission from our own galaxy and in the process found a tiny amount of radio noise that they could not explain. But they doggedly pursued the source of the noise—even evicting a pair of pigeons that had inhabited and dirtied their antenna—and concluded that they could not explain the noise by any known phenomenon. The cosmic connection was established with help from the astrophysics gossip network Penzias learned from American radioastronomer Bernard Burke that Ken Turner of Johns Hopkins University had heard a talk by P.J.E. Peebles of Canada in which he said that based upon his calculations there should be radiation left over from the big bang. Peebles, unaware of the earlier work by Gamow and collaborators, had duplicated their calculations. Moreover, Peebles' Princeton colleagues, Robert Dicke, P.G. Roll, and David T. Wilkinson, had just set up an experiment to search for the radiation that Penzias and Wilson had just found! The discovery was announced in a paper by Penzias and Wilson in the Astrophysical Journal, modestly entitled "A measurement of excess antenna temperature at 4080 MHz," followed by a paper by Dicke, Peebles, Roll and Wilkinson giving the big-bang interpretation. In 1978, Penzias and Wilson were awarded the Nobel Prize in physics.

in the universe, we do know that: (1) most of the matter is dark (that is, does not emit detectable radiation); (2) the theory of big-bang nucleosynthesis implies ordinary matter contributes between about 1 percent and 15 percent of critical density; and (3) several measurements indicate that the total mass density is well in excess of 15 percent of critical and perhaps close to critical. (If the density is greater than the critical density gravity eventually halts the expansion and the universe recollapses; if the density is less than the critical density the universe expands forever.)

If, as seems likely, the mass density is greater than 15 percent of the critical density, then there must be another form of matter in the universe! The best prospect is elementary particles left over from the early, fiery moments. Theories that unify the forces of nature predict the existence of new particles, and for several of them (the axion, the neutralino and the massive neutrino) their expected mass density today is close to critical density.

▲
THE TRAVEL DISTANCE
TO THE ANDROMEDA
GALAXY IS OVER
1 MILLION LIGHT-
YEARS. THE CLOSEST
BRIGHT RADIO GALAXY,
CYGNUS A, IS ALMOST
1 BILLION LIGHT-
YEARS AWAY FROM
EARTH.

Knowing the composition of dark matter and the nature of lumpiness allows the evolution of structure to be simulated on a computer and compared to what is seen. The most promising theory is motivated by inflation and is known as cold dark matter. It asserts that lumpiness arose during inflation and that the matter is composed of 5 percent critical density in ordinary matter and 95 percent of critical density in slowly moving relic particles (axions or neutralinos). Thanks to technological advances (CCD detectors, new large ground-based telescopes, and space-based telescopes) cosmologists are putting the cold dark matter theory to the test, measuring the small variations in the temperature of the CBR, studying the distribution of galaxies, searching for the most distant (oldest) objects we can see, and even trying to detect the relic axions or neutralinos in our own galaxy.

Whether or not cold dark matter becomes part of the grander big-bang theory remains to be seen. If it does, the study of structure formation will have provided a unique window on the earliest history of the universe and thereby on the unification of the fundamental particles and forces of nature.

DISAPPEARING SPECIES

As fragments of comet Shoemaker-Levy 9 slammed into Jupiter in July 1994, one newspaper cartoon showed worried Jovian dinosaurs looking at the sky, saying, "Not again!" No, there are no dinosaurs on Jupiter. The cartoon simply

Stuart L. Pimm

reminded us that on several occasions in Earth's geological history, large numbers of species have become extinct, perhaps because of the impact of comets and asteroids. Ecologists believe that Earth is currently experiencing an extinction catastrophe as large as the ancient geological ones. But in this case, the catastrophe will not be a result of a comet or asteroid.

Instead, ecologists point the accusing finger at our own species—the problem is our success. From Africa, we have spread over Europe and Asia, then Australia, into the Americas (probably within the last 10,000 years or so), and, even more recently, across the remote islands of the Pacific. We are one of the most abundant species on the planet, and our numbers are increasing rapidly. Each year we consume about 40 percent of total plant growth on land. (Some of this consumption is what we and our domestic animals eat; some is the wood we burn or use for other purposes.) In the oceans, we have repeatedly taken out more than the annual growth—the

WHICH INSECT WILL BECOME EXTINCT NEXT?

"interest" in our ecological savings account—and the balance has shrunken accordingly. Many once-profitable fisheries, such as herring and cod, are now virtually depleted.

How do ecologists estimate the number of victims of this catastrophe? The obvious answer is to look at our world catalog of animals and plants and to place a mark against each species when it becomes extinct. Unfortunately, we do not possess such a catalog. We do not even know how large it is. Astronomers' estimates of the Hubble constant (it determines the size of the universe) differ by a factor of approximately two. Biologists' estimates of the total number of species differ by a factor of 20 or more. Only for groups like birds do we have a nearly complete catalog. For other groups like insects, bacteria, and fungi, we only know that we do not know most of the species. How do we know how many are being lost?

In order to estimate the number, we use indirect methods. First, we can guess the fraction of well-known species that are being lost. If these are typical of species

◄

ALL THAT REMAINS OF THE DODO ARE INCOMPLETE SKELETONS, TWO HEADS AND TWO FEET IN VARIOUS MUSEUMS IN EUROPE, THE UNTIED STATES, AND MAURITIUS.

we do not know, we can multiply. We know that there are about 10,000 species of birds. Suppose there have been a thousand recent extinctions, or 10 percent of the total. If there are two million animal and plant species, then we might have lost 200,000; if there are 50 million species, then we might have lost five million. These are rough guesses. (I have yet to justify the figure of a thousand recent bird extinctions.) Scientists frequently make estimates to the nearest factor of ten. Let me make one more: how many extinctions should we see if our species wasn't guilty?

From the fossil record, we know that most species become extinct. We can also estimate—to the nearest factor of ten—how long a species lasts. A typical answer is 1-10 million years. If we watch a million species for somewhere between a year and a decade, we should see an extinction. If we watched 10,000 bird species, we should see an extinction once a century only (perhaps even once a millennium only). Even if we criss-crossed the planet annually, a bird extinction should be a once-in-a-lifetime event. Instead, however, dozens of species are dying, and many more near death.

In order to prove that humans are the principal destroying factor in these extinctions, we need to look at those places where the first human presence is most recent. On Pacific islands the evidence is everywhere. In Hawaii, the last male 'O'o'a'a' sang in a remote swamp for years without finding a mate before he, too, died. Searches for the "O'u and the Nukupu'u have gone unrewarded, and year-long searches for the Po'o uli have found only scattered individuals. On Guam, an accidentally introduced snake species exterminated all the island's forest birds.

WHICH REPTILE?

Evidence mounts as we go further back. Digging the dirt of Hawaiian cave floors, scientists have found bones of dozens of bird species that failed to survive Polynesian colonization a thousand years ago. Other species might have disappeared without a trace. There is a simple way to estimate how many had previously existed. We classify species in two ways—first, those for which we have specimens in our museums ("skins") and those for which we do not ("no skins"), and second, those for which we have bones ("bones") and those for which we do not ("no bones"). Combine these and there are four groups. One is "no bones and no skins"—the missing species. An estimate of the number of these species is "bones but no skins" multiplied by "skins but no bones" divided by "bones plus skins."

Now the body count. Add the estimates of missing species to those we know only from their bones, then add those to the species that have become extinct since ornithologists first visited the islands. For the Hawaiian islands, the total is more than one hundred. Across all the islands in the Pacific the total body count is about a thousand, including the giant moas of New Zealand. Some scientists who collect these bones think the correct number will be nearer two thousand. Even if these were the only bird extinctions—that is, if the rest of the planet were untouched—they would represent a loss of 10-15 percent of all bird species. Losses elsewhere in the world—and there are plenty of them—add to the total.

WHICH PLANT?

There is nothing particularly unusual about bird extinctions, except that they are easy for us to record. Of about a thousand Hawaiian plant species, approximately one hundred have become extinct since collecting first began, and a further hundred species are very near extinction. How many plant extinctions the Polynesians caused is unknown, as are the extinctions of insects and other animals that depended on these plants.

Are extinctions still continuing? To answer this question, we must use the second indirect method. We know that the destruction of habitat is often the most important cause of extinction. (Accidentally introduced species, like the snake on Guam, and the secondary extinctions of, say, insects when their food plants are lost, are also causes.) Consider the tropical forests that are burned to clear land for agriculture. Will this clearing cause extinctions?

Ecologists use a rough-and-ready relationship to estimate the loss of species from the loss of habitat. For example, if we lose all but 1/16 of the forest, then we predict a loss of half the species. We can only test the prediction where we know the forest losses and species losses in some detail. One such place is the tropical forests of the Philippines south to New Guinea and west to Sumatra. Satellite images show that much of the forest has already been cleared and ground surveys show that many species of birds are threatened with extinction as a consequence. Applying such calculations to tropical forest losses worldwide suggests we will soon be losing many species of animals and plants.

Given that our population will at least double, we will continue to lose species, and probably at an accelerating rate. We will lose more than just species, for it is species that are the basis of our modern medicines. They also contain the genes that improve crops, and that in total, provide the ecosystem services that keep our planet healthy and our atmosphere low in carbon dioxide.

SENSE FROM NONSENSE

Over the last century and a half, investigators have debated whether the human organ is best conceived as a whole or if it can be thought of as consisting of separate areas, each responsible for specific functions. Progress in brain research, its most impressive insights and achievements, can best be understood within the context of this continuing debate.

In 1848, an explosion at a work site on the Rutland and Burlington Railroad in Vermont shot an iron rod through the head of Phineas Gage. While he survived the accident, Gage suffered grievous destruction to the frontal lobes of his brain. As a result, he went from a conscientious, hard-working man to someone who could no longer be depended upon. His personality change, resulting from frontal-lobe injury, supported the then-emerging concept that different parts of the brain serve different functions.

Richard Restak

Additional proof for the localization theory came in 1861, when the French neurologist Paul Broca demonstrated the site in the brain responsible for loss of speech. A patient under Broca's charge had suffered a stroke that had deprived him of speaking anything other than the word "tan." Twelve years later, the neurologist Carl Wernicke discovered a second area of the brain that, when damaged, resulted in a different speech disturbance. These discoveries served to establish neurology as a scientific discipline aimed at correlating brain injuries and losses of function.

In 1870, Gustav Fritsch and Julius Eduard Hitzig, both of Germany, carried out experiments on Hitzig's kitchen table showing that discrete movements of the face or limbs of a dog could be elicited by electrically stimulating the dog's frontal cortex on the side of the brain opposite the limb movements. Electricity then emerged as a major research tool.

In 1876, David Ferrier, a British researcher, refined electrical techniques and extended their use to monkeys. Soon surgeons began applying the knowledge to humans and operating on the basis of functional maps of the brain. In turn, surgical cases confirmed that the brain can be functionally organized into distinct areas, culminating 80 years later with operating-room experiments performed by Canadian neurosurgeon Wilder Penfield. By electrically stimulating the exposed brains of his cooperative patients, he constructed maps of human sensory and motor functions. Penfield was the creator of cartoon-like drawings of body parts,

◄

SELF-PORTRAIT PAINTED BY A SCHIZOPHRENIC PATIENT

arranged within the brain, not according to their actual size, but according to their importance in the sensations and motor activities of daily life. These homunculi were drawn with enormous fingers, tongue and lips, while the knee and the entire back appear tiny in comparison.

In 1871, Camillo Golgi of Italy developed a method for staining nerve cells with silver nitrate, but incorrectly concluded from this research that the fine filaments of the brain formed a unified network, or syncytium. In 1900, the

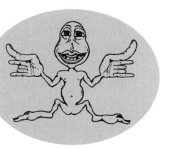

▲

HOMUNCULUS
("LITTLE MAN"), A
TRADITIONAL WAY
OF DEPICTING HOW
THE BODY IS LIKELY
TO BE REPRESENTED
IN THE SENSORY
PARTS OF THE BRAIN.

Spanish artist-turned-physician, Ramon y Cajal, perfected Golgi's technique and correctly concluded, on the basis of his own staining work, that each neuron is separated from others by a junction, the synapse, named in 1897 by the English neurophysiologist Sir Charles Sherrington. But the insight left investigators with a problem: how is the nerve impulse conducted across the synapse? The answer came in 1921 in an experiment suggested in a dream to German physiologist Otto Loewi. His identification of the neurotransmitter acetylcholine spurred additional efforts, which continue to the present, to identify all neurotransmitters, the chemical messengers by which brain cells communicate across the synapse.

In the 1920s, Walter Cannon and Walter Hess, working independently, established the importance of the hypothalamus as a regulator of the body's physiological processes and a mediator of some aspects of emotional expression.

Hans Berger of Germany showed, in 1929, that rhythmic electrical activity can be recorded from the surface of the human brain. The new science of electroencephalography would, half a century later, employ computers in so-called "evoked potential studies." Using this method, brain-wave patterns, resulting from specific stimulations, such as a flash of light or a sound, can be isolated from the brain's background electrical activity.

For instance, brain activity can be measured prior to one's conscious intention to act, a finding that suggests consciousness may best be understood as an emergent property of the brain when it reaches a threshold of complexity. Most neuroscientists believe that only the human brain is sufficiently complex to support consciousness, but the issue cannot be settled definitively because consciousness is inseparable from the sophisticated language capabilities that we alone possess.

In 1937, James Papez, a neurologist at Cornell University, proposed a new circuit to account for emotions. Twelve years later, Paul MacLean of Yale University enlarged upon Papez's concept and coined the term "limbic system," so named because all of the structures comprising it completely surround the lower threshold of each cerebral hemisphere, forming a circular edge, or border (from the

Latin limbus). In the 1940s, Guiseppe Moruzzi and Horace Magoun discovered the brainstem reticular formation and its role in wakefulness, further confirming the importance of subcortical structures in awareness and consciousness. In 1949, the Canadian neuropsychologist Donald Hebb proposed that neurons form operational networks, strengthened or weakened according to patterns of use. With each repetitive act, a network is facilitated—a unifying concept of brain activity from neurons to behaviors.

In the 1950s and 1960s, physiologists David Hubel and Torsten Wiesel, at Harvard University, and Vernon Mountcastle, at Johns Hopkins, showed that the cerebral cortex is arranged in a columnar manner, with dedicated neurons responsive to specific environments. In the visual cortex, for instance, some cells respond to lines of certain orientation, while unresponsive to lines of only a slightly different orientation. The introduction of tranquilizers and antidepressants supported the notion of a biological basis for all behavior. Subsequent research suggested that specific neurotransmitters were involved in certain forms of psychoses and depression. Today, it is no longer believed that such disturbances in normal thinking and feeling can be accounted for by single neurotransmitter abnormalities. More likely, many neurotransmitters are involved in specific disorders.

In the 1960s, California Institute of Technology neurobiologist Roger Sperry further advanced the theory of localization by working with patients who had undergone surgical disconnection of the two hemispheres of the brain. Sperry demonstrated that certain functions are related to one or the other side of the brain. In the 1970s, computers linked with X-rays resulted in the introduction of CAT scans, and this technique was soon followed by MRI, magnetic resonance imaging. Both techniques provide images of brain structure but tell us nothing about what is happening in the brain and how many brain areas are involved.

When PET scans were introduced in the 1980s, the long-dormant theory of holism was resurrected. Multiple, and often widely separated, areas of the brain are illuminated on PET scans, as subjects engage in straightforward activities like looking, listening, or speaking. These diffusely distributed networks not only confirm what Hebb believed, but have reintroduced holistic ideas, largely abandoned in the nineteenth century.

Today, the neurosciences have entered the era of molecular biology with the discovery of genes responsible for certain brain diseases, like Huntington's chorea and certain instances of inherited manic depression. The goal now is to develop ways either to influence genes by some form of gene therapy or to discover ways to reverse the harmful effects programmed by the genes, by means of drugs.

X-RAY COMPUTED
TOMOGRAPHY (CT)
OF BRAIN

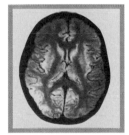

MAGNETIC RESONANCE
IMAGE (MRI) OF BRAIN

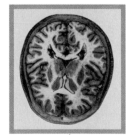

BRAIN SLICE

SON ET LUMINAIRE

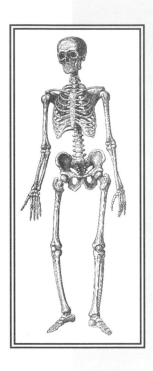

◄

DARWIN USED HUMAN ANATOMY AS EVIDENCE IN HIS ARGUMENTS THAT HUMANS EVOLVED FROM ANIMALS.

72

BRITISH ARCHAEOLO-GIST GEORGE SMITH reads tablets sent to him from the excavation of a library in the ancient ruins of Nineveh, a city destroyed about 660 B.C. He is startled to find a report of a great flood much like that described in the Bible. Smith has uncovered the story of Gilgamesh, humanity's oldest surviving literary product.

73

GERMAN PSYCHOLO-GIST WILLIAM WUNDT proposes that facets of human behavior can be measured, founding the field of experimental psychology. He publishes *Principles of Physiological Psychology*—considered one of the most important books in the history of psychology.

CABLE CARS, the world's first street cars, begin operating in San Francisco.

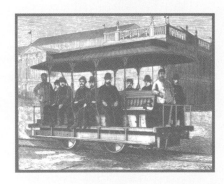

1871

WHILE HE HAD DELIBERATELY avoided controversy by ignoring human evolution in *On the Origin of Species*, Charles Darwin now publishes *The Descent of Man*, which suggests that humans evolved from animals. His evidence includes humans' vestigial tailbone and apparently nonfunctional organs, such as the appendix. No fossils of apelike humans or human-like apes are yet available to buttress Darwin's theory.

75

GERMAN EMBRYOLO-GIST OSKAR HARTWIG, studying the reproduction of sea urchins, becomes the first person to witness the union of an egg cell and a sperm cell. He notices that despite the abundance of sperm cells surrounding the egg, only one successfully enters it.

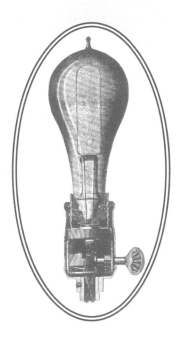

76

HUMAN COMMUNICA-TION IS CHANGED forever when American inventor Alexander Graham Bell spills battery acid on his pants and calls out to his assistant, "Watson, please come here; I want you." Working on a different floor, Thomas Watson, hears Bell's cry through the device they are working on, which comes to be known as the telephone. Bell patents his invention a few months later, and within two years the first telephone exchange is established in New Haven, Conn.

77

U.S. INVENTOR THOMAS EDISON, who a year earlier established the world's first industrial research laboratory in Menlo Park, N.J., gives the facility instant recognition by inventing the phonograph.

79

THOMAS EDISON LETS THERE BE LIGHT when he sends a current through a carbon wire filament in an evacuated glass bulb. The white-hot filament lights the bulb and burns continuously for 40 hours. Edison patents his light bulb and on New Year's Eve, before a crowd of 3,000, floods the streets of Menlo Park with electric light.

U.S. CHEMIST IRA REMSEN and his student, Constantine Fahlberg, synthesize an organic compound named orthobenzoyl sulfimide. Fahlberg mistakenly touches his mouth with a finger that carries a few grains of the compound and is surprised by the sweet taste. The compound comes to be known as saccharin.

◄

IN ADDITION TO
THE PHONOGRAPH,
EDISON HELD 1,092
OTHER PATENTS—
A WORLD RECORD.

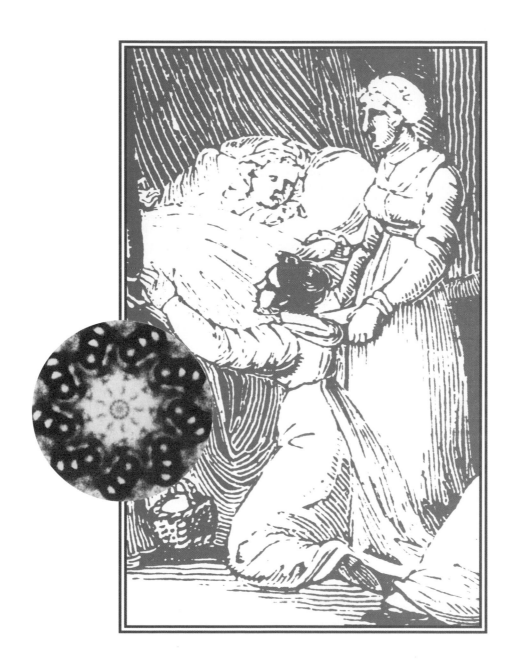

AN ANCIENT LEGACY

For half a century after *Scientific American* was founded, most people thought cancer was a contagious disease, a growth of foreign matter that somehow took root in an otherwise healthy body. The first cancer hospitals functioned more as quarantine units than specialized care facilities, and those families affected by cancer went to great lengths to hide the condition from the outside world.

Robert A. Weinberg

Far fewer succumbed to the disease in 1845 than today. The first cancer epidemiologies, published in the early 1840s in Paris and Verona, showed that less than 2-$\frac{1}{2}$ percent of deaths could be attributed to malignancies. In fact, few lived long enough to contract the disease which, then and now, affects mostly the elderly. A century later, one in six died from cancer. What had once been a rare disease now became commonplace.

The world of cancer research underwent a revolution in the 1840s. For the first time, the microscope was used intensively to study the invisible world of normal and malignant tissues. By the beginning of the decade, German pathologists had determined that normal tissues are built of discrete units—the cells each of which has a life of its own. By 1855, the greatest of these pathologists, Rudolf Virchow, expounded his dictum that *omnis cellula e cellula*—all cells arise from other cells. If Virchow was right, as indeed he proved to be, then all the cells in the body must descend, through repeated rounds of growth and division, from the giant ancestral cell, the fertilized egg.

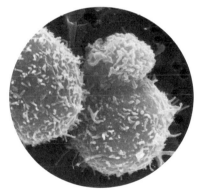

Virchow's vision was soon applied to cancer. The origins of cells in a tumor were traced to cells present in normal tissues. Another century passed before this idea was refined further—all the cells in a tumor descend from a single common ancestor, a renegade that begins to proliferate uncontrollably. Through continued growth and division, the renegade spawns millions, then billions of descendants that together aggregate to form a tumor visible to the naked eye. Cancer was no longer thought to be an alien invader. Rather, malignant growths could only be viewed as normal, out of control tissue.

Until the 1960s, almost all cancer research was performed studying tumors with the naked eye or under the microscope. But if the roots of a tumor were traceable to a single runaway cell, then key questions about cancer's origins turned on the sources of this cell's behavior. The naked eye and the microscope

◀ ▲

BY DIRECTING THE ASSEMBLY OF A CELL'S SKELETON, CENTRO-SOMES *(LEFT)* CONTROL DIVISION. RENEGADE DIVISION CAN FORM MALIGNANT GROWTHS. THE BODY CAN REACT BY SENDING LYMPHO-CYTES TO BIND WITH TUMOR CELLS *(ABOVE)*.

offered no help here. Questions about cell behavior could only be addressed by examining the complex soup of biochemicals inside the single living cell.

Over the last two decades, modern cancer research has been further alienated from its historical antecedents. In recent years, many of the solutions to understanding cancer's origins have come from scientists whose research interests are far removed from human disease.

The Bread of Life

Cells maintain an elaborate apparatus to ward off the chaos brought on by damaged genes. This mechanism monitors the cell's DNA molecules, looking for lesions in the double helix—a tear here, a gap there, or an inappropriately inserted base somewhere else. The cell's DNA repair apparatus detects flaws and erases them. Without repair, mutations rapidly accumulate.

Sooner or later, mutations damage growth-regulating genes, and cancer ensues.

Richard Kolodner, working at the Dana-Farber Cancer Institute in Boston, has spent the last decade studying DNA repair genes in the baker's yeast commonly used to make bread rise. A cancer institute would seem to be an unlikely place to study yeast and its genes, but cancer researchers have learned, time and time again, that the clues to cancer turn up in unexpected places.

Kolodner found two genes in the yeast cell that are responsible for DNA repair. When either of these genes is damaged, the yeast cell rapidly accumulates widespread mutations throughout its genetic blueprint.

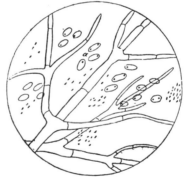

YEAST PLANT IMAGE FROM THE 1800s.

Kolodner realized that similar widespread mutations were uncovered in human cancer cells by Manuel Perucho, who works in LaJolla, California. So Kolodner sought the human counterparts of his yeast DNA repair genes. Though more than a billion years removed from one another, the human genes that he found still bore striking similarities to their yeast counterparts.

This discovery quickly led to a new understanding of one kind of human cancer. Damaged versions of one or the other of these two human DNA repair genes are now known to be often passed in human families, where they lead to a high risk of colon cancer. The colon cells of afflicted individuals, lacking the ability to repair their DNA effectively, accumulate widespread mutations, some of them affecting critical growth-regulating genes. Soon these genetically damaged cells begin uncontrolled, cancerous growth. Study of the obscure and seemingly irrelevant—tiny baker's yeast—led straight to the heart of the human cancer problem.

This conundrum makes little sense until you remember the lesson taught by Charles Darwin. The design and function of our body is of ancient lineage. Our pedigree is traceable to relatives like apes, then to primitive mammals, and before them to reptiles, amphibians, fish, and, ultimately, a billion and more years ago, to simple, single-cell organisms. Our brain may be rather novel, but almost everything else about us is standard, off-the-shelf biochemical and cellular hardware that was introduced a long time ago.

Our cellular hardware is subject to breakdowns similar to those occurring in the cells of our distant relatives. Cancer occurs in our biological cousins 50 and even 500 million years apart from us on the evolutionary tree. Studying cancer cells of these relatives often tell us a great deal about our own.

The striking similarity of our cellular machinery and its occasional flaws has only become apparent in the last decade when we were able to study the components of this machinery—macromolecules like DNA, RNA, and protein—up close. Only recently have we begun to realize that the molecular machinery inside our cells was developed more than a billion years ago. Once optimized, the plan, spelled out in our genetic blueprint, was passed in almost unchanged form through the countless generations that separate us from our very distant and simple ancestors. We now know that each individual cell within a complex organism like the human body relies on its ancient complement of genes to assemble its internal machinery and to program its growth.

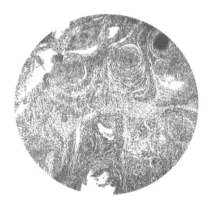

Over the past decade, we have learned that a renegade cancer cell grows abnormally because it carries damaged, mutant versions of the ancient growth-programming genes that constitute the common evolutionary inheritance of all animal cells on this planet. Seeking instruction about when it should grow or remain quiet, the cancer cell encounters misleading information in its damaged genetic blueprint. The bad genes—mutant oncogenes and tumor suppressor genes—lead directly to abnormal growth.

We have learned about the blueprint by studying flies, worms, yeasts, sea urchins, clams, birds, and, on occasions, even our close mammalian relatives such as mice and rats. We have even learned about these genes and the ways in which they operate by studying human cancer cells, but only on occasion. Increasingly, solutions to the human cancer problem come from those who snoop around in obscure corners of the biological world and uncover clues that tell us how damaged genes and aberrantly functioning proteins push the renegade cell into the runaway growth that leads to the chaos of cancer.

▲

IN 1887, *SCIENTIFIC AMERICAN* REPORTED ON A "BENIGN" GROWTH *(ABOVE)* IN THE THROAT OF CROWN PRINCE FREDERICK OF GERMANY. IN 1888, HE DIED OF THROAT CANCER, 99 DAYS AFTER ASCENDING THE THRONE.

RHYTHM OF THE EARTH

Shortly before *Scientific American* began publication in 1845, a young Swiss naturalist, Louis Agassiz, examined scratched rocks found in Swiss valleys and boulders mysteriously removed from their place of origin. These observations led Agassiz to introduce an important and surprising hypothesis. Over the next twenty years, he convinced the small community of natural scientists that ice had

John Firor

covered much of Europe thousands of years ago and was responsible for these geological curiosities. Many others had already come to the conclusion that glaciers had once reached further down the valleys.

But Agassiz's evidence of an extended and widespread much colder climate—an ice age—accelerated the study of early climates and encouraged efforts to understand the structure of today's climate.

Soon Agassiz and others realized that there had been a series of glaciations, each one erasing much of the evidence of the earlier glaciers, but leaving a few clues that could help establish the sequence. Then came the realization that the ice ages in Europe, America, and even in parts of the Southern Hemisphere, had occurred at the same time—that the emergence and receding of ice was a global phenomenon. Much more recently, scientists discovered that the ebb and flow occurs with regular frequency; the rhythm of ice ages is set by small changes in the earth's orbit around the sun as well as variations in the tilt of the earth's axis.

Before the role of orbit changes was confirmed, however, many proposals were introduced to explain these cycles. *Scientific American* reported several of them—in 1952, Charles Warren credited mountain-building episodes for creating periods of repeated glaciations and, in that same year, Harry Wexler thought volcanic explosions played an important role. In 1958, Ernst Öpik suggested a scheme in which changes in the internal workings of the sun were manifest as ice ages on earth. None of these theories fully explained the observations made, but in hindsight each scientist had asked important questions about climate, and each process is still being studied. In 1948, George Gamow, already famous both as a physicist and a popular science writer, presented an excellent description of the astronomical theory, including a clear explanation why the earth's orbit should slowly vary. While Gamow's ideas proved to be the most far-sighted, his imagination went beyond what is now thought to be the cycle of glaciation. Gamow conjectured that 20,000 years from now, "No trace of ice will then be left in the northern regions, and the

◄

CLIMATIC PATTERNS
ARE VERY EVIDENT IN
THE GROWING AND
SHRINKING OF THE ICE
FIELDS OF PATAGONIA.

shores of Baffin Bay will be covered with palm groves, their leaves rustling softly under the caressing breath of the northern winds."

Half a century earlier, the Swedish scientist, Svante Arrhenius, considered the role of gases in the atmosphere—primarily water vapor and carbon dioxide— that help keep the earth warm. If carbon dioxide were lacking, he wondered, could that have been responsible for the cold periods. Calculating the earth's temperature, if the atmosphere possessed only half its current carbon dioxide, he concluded that the earth's surface would be cooler by several degrees, enough to account for an ice age. The American geologist, T.C. Chamberlin, proposed similar ideas, briefly described in *Scientific American* in 1907. But neither Chamberlin nor

► THE RHÔNE GLACIER HAS BEEN RETREATING FOR MORE THAN 250 YEARS AS SHOWN BY AN 1848 WATERCOLOR BY HENRI HOGARD *(TOP)* AND A PHOTOGRAPH FROM 1970 *(BOTTOM).*

Arrhenius could satisfactorily explain how the air could shed, and then regain, half its carbon dioxide.

A few years later, however, as Arrhenius looked out of his office window and saw smokestacks "evaporating our coal mines into the atmosphere," he understood how the level of carbon dioxide could change. He then calculated the effects of doubling the amount of carbon dioxide in the air and concluded that it would raise the earth's average temperature by several degrees, about the size of an ice age cooling but in the opposite direction. Today, hundreds of scientists worldwide are attempting to improve the calculation of a possible human-induced climate heating while others attempt to estimate the impacts such a sharp climate change would have on crops, ecosystems, and human activities. A few scientists remain skeptical that people can have such a profound influence on climate, and a few others cling to the belief that such a change would be beneficial.

The work of Agassiz and Arrhenius have merged in several ways. For those who follow Arrhenius's lead, studying the influence of carbon dioxide, orbit variations test their theories. Ice-age theorists and climate-change scientists are increasingly working together to explain how small changes in the earth's orbit can produce profound climatic shifts.

Measurements derived from polar ice sheet samples present a record of temperatures over more than 160,000 years. This record shows the changes from a long ice age to a briefer warm interval, back to another long, cold period, and on to our present warm climate. Fossils and pollen preserved in bogs and elsewhere supplement data gathered from ice sheets, indicating that ancient climate changes disrupted ecosystems. In this century, the retreat of glaciers observed in many locations suggests that our climate has became warmer. With continuing carbon dioxide emissions, one wonders which parts of Gamow's vision might not be fanciful.

Considering the more distant past, in particular, the Cretaceous period more than 65 million years ago—with its luxurious plant growth, dinosaurs, and much warmer temperatures than now—climatologists attempted to explain the different climate by the effect of continents, in their slow drift, being in different locations than they are today. But the changed positions were not enough; these scientists found it necessary to assume that the Cretaceous atmosphere may have contained several times as much carbon dioxide as we have today.

Ice Detectives

The search for the mysterious cause of ice ages is a story that rivals fiction. Analysis of polar ice found the water in these samples slightly different— lighter—than ocean water, the differences arising from two slightly different forms of oxygen. As ice caps build therefore, removing light water from the oceans, ocean water should become a bit heavier. Elsewhere, ocean biologists discovered that the bodies of small creatures living deep in the ocean accurately preserved the oxygen composition of the ocean water. So it seemed possible that the shells of these creatures, deposited in layers on the ocean floor over a million years, preserve a record of the amount of ice stored on land.

But could this record be deciphered? Cores of sediments were brought up from the bottom of oceans and examined for the remains of small creatures

contained in them. Some cores showed large accumulations of sediments each century, but these cores did not cover a sufficient number of centuries. Others had small accumulations—as little as one millimeter per century, for example—but in these cores the disturbance by living creatures mixed sediment materials from different times. Exact dates of the older layers in the cores were also uncertain. Finally, in 1976, two cores from the bottom of the Indian Ocean were found to have the right accumulation rate. Using these cores to perform studies of the earth's magnetic field, it was shown that the field had reversed some 700,000 years ago, giving a faint change in the core and solving the dating problem. These two cores showed clearly the characteristic variations of cycles in the earth's orbit and placed the astronomical theory of ice ages on a firm basis.

◄

ICE CORES ALSO CAN AID IN THE INVESTIGA-TION OF CLIMATIC CHANGE. ANALYSIS OF GASES AND HYDROGEN RATIOS IN TRAPPED BUBBLES OF ANCIENT AIR HELPS SCIENTISTS RECONSTRUCT A 160,000-YEAR-OLD RECORD OF TRACE GASES AND TEMPERATURES.

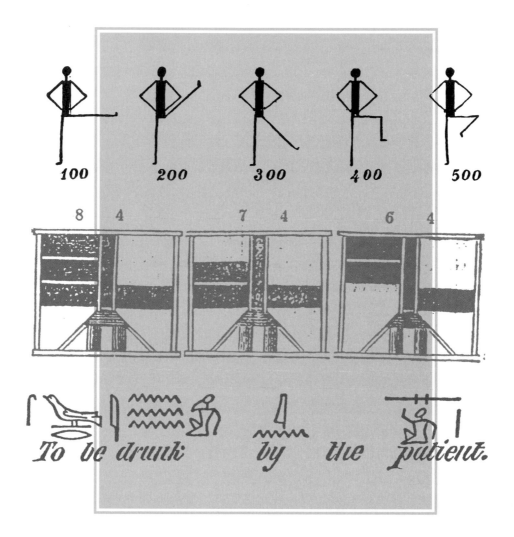

SENDING SECRET MESSAGES

Cryptography is the study of codes and ciphers to ensure the privacy and authenticity of information. While most people think of cryptography primarily in terms of privacy, in modern applications authentication is at least as important. Considering electronic banking: is it worse if someone can violate your privacy by reading checks you write, or violate authenticity by forging checks on your account?

Martin E. Hellman

A simple substitution cipher (e.g., A=Z, B=X, C=B, D=V) can be broken too easily to be of practical value, but it is helpful in understanding how more secure cipher systems work. The scrambled alphabet, or key, is known to authorized users so that they can easily encipher and decipher all messages flowing between them. Anyone who does not know the key has difficulty in deciphering intercepted messages, thus violating privacy, or in enciphering messages of his choosing, thus violating authenticity.

Since most modern ciphers use the same mathematical operations to encipher and decipher, it is as difficult to encipher without the key as it is to decipher. Such systems provide authentication as well as privacy for users. Modern ciphers are also much more secure than this simple substitution, making it impossible for an opponent to violate either privacy or authenticity.

Cryptography experienced major growth during the First World War because of the advent of radio. No wires needed to be strung to communicate with a distant friendly unit, but no wires needed to be tapped by a distant enemy unit to listen in either. Satellite communications have grown spectacularly in recent years, and a great deal of telephone traffic is carried by microwave radio. Cellular telephones are a recent development. Because so much of today's communication is carried by wireless means, most people whose calls are tapped never discover the intrusion.

The transmission of computer-readable messages, such as electronic mail, increases the need for encryption, since over 10 billion words can be searched at a total cost, including equipment usage and power, of just one dollar. This makes it cost effective for an industrial spy to search through millions of messages and record only those which contain key words, such as "IBM" and "research."

Modern cryptography owes much to the pioneers of the late-nineteenth and early-twentieth centuries, notably Prussian military officer Friedrich Kasiski, Dutch scholar Auguste Kerckhoffs, Gilbert Vernam, William Friedman, and Claude

◄

WITHOUT THE NECESSARY KEYS, THE INFORMATION CONTAINED IN THESE MESSAGES IS RENDERED USELESS.

Shannon. But the computer revolution was necessary before their ideas for better systems could emerge. Because encryption is a special-purpose form of computation, as long as good computers were too costly, so were good encryptors. The primary computing device in the Second World War was the gear-based adding machine.

Until the advent of the Data Encryption Standard (DES), promulgated by the National Bureau of Standards (NBS) in 1977, the commercial encryption market remained small. DES was developed by IBM, which initiated a cryptographic research effort in 1968. DES could be implemented on a single integrated circuit, at a cost ranging from $5 to $100. Today's integrated circuits are one hundred times as complex as those of 1977, so DES occupies only a small fraction of a modern chip and the cost is negligible.

Because of its low cost and status as a standard, DES became the world's most widely used system. It also became the center of controversy. Immediately after DES was introduced, some cryptographers argued that its 56-bit key was too small, allowing "only" 72,000 million million keys—a number they claimed could be searched at a cost of $10,000 on a special-purpose spying machine. NBS and its ally in creating and defending DES, the super-secret National Security Agency (NSA), disputed the claim, and this question still generates controversy. A quick fix was proposed: use DES three times, with three different keys. Although this only increases the cost of encryption by a factor of three, recovering one key by exhaustive search costs more than the next million years' gross national product.

The controversy was complicated because one of NSA's major missions was spying on foreign communications, many of which were unencrypted, but would be encrypted in DES as it became popular. Making DES unbreakable would thus put a major division of NSA out of business. More recently, the Federal Bureau of Investigation has joined the controversy because criminals and terrorists using encryption are beginning to foil its court-ordered wiretaps.

Key distribution—transmitting keys to legitimate users, while keeping them secret from opponents—is a major problem. Couriers were used traditionally, but they are too slow and expensive for widespread commercial use. In 1976, Whit Diffie, Martin Hellman, and Ralph Merkle, all of Stanford University, proposed public key cryptography as a solution. Each user is given a pair of inverse keys, one public, and one private. When B wants to communicate privately with A, he looks up A's public key, encrypts the message with it, and transmits the encrypted message. Only A knows the inverse key (A's private key) which can decrypt the message. Couriers are not needed because the public keys needed to encrypt can be published in an electronic "telephone book."

Public key systems offer another new possibility, digital signatures. Conventional systems provide authentication against third-party forgeries but cannot settle disputes as to what message, if any, was sent. Because the same secret key is used by the transmitter and the receiver, anything the transmitter can do to "sign" a message can also be performed by the intended receiver—but not by a third party. In contrast, a written signature can be checked by anyone, but ideally only generated by one person. A public key system produces a digital signature, if the signer's private key is used to scramble the message, in which case his public key is used to verify it.

Digital signatures promise to revolutionize electronic business but have been stymied by the creation of standards. Under pressure from national security and law enforcement interests, the National Institute of Standards and Technology (NIST, formerly NBS) delayed the introduction of a public key standard. When NIST finally proposed a Digital Signature Standard (DSS) in 1991, based on work of Taher ElGamal, a PhD student at Stanford, and Claus Schnorr, professor of mathematics at Dee Johann Wolfgang Goethe Universitat in Frankfurt, it covered signatures, but not privacy, because signatures do not threaten national security or law enforcement.

Significant controversy now surrounds DSS, with most opponents preferring a public key system, permitting privacy and signatures. One such system, RSA, invented in 1977 by Ronald Rivest, Adi Shamir, and Leonard Adleman, at MIT, depends on the difficulty of factoring large integers for its security. RSA was in widespread commercial use prior to the announcement of DSS, and it is not clear how the contest will end.

The Clipper-Chip Controversy

In response to National Security Agency and Federal Bureau of Investigation concerns, in 1993, the Clinton Administration announced an encryption initiative known as "the Clipper chip," or "key escrow." According to this plan, each chip has a different master key, held in "escrow" by a governmental entity. When a wiretap is ordered by the courts, the targeted chip's master key is turned over to the requesting law-enforcement or national security agency, which can then unscramble the encrypted communications. In an ensuing controversy, one side warns of lawlessness and terrorism and the other of "Big Brother" intrusion. Obviously, when government can spy on the "bad guys," it can also spy on "good guys."

▲

THE HOLLERITH CARD-PUNCHER

1880

WITH THE U.S. CENSUS FACING THE TASK of sorting through increasingly massive amounts of information, mechanical engineer and bureau employee Herman Hollerith seeks a solution. He develops a machine that turns information on a punch card into readings on a dial that a person can easily record. The first electromechanical calculator launches Hollerith on a career that culminates in his founding of the Tabulating Machine Company, which later becomes IBM.

FRENCH PHYSICIAN Charles-Louis-Alphonse Laveran isolates the microorganism that cause malaria. A protozoan, it is the first pathogenic organism to be found that is not a bacterium.

81

SEEKING A CURE FOR ANTHRAX—a deadly disease that affects humans and domesticated animals—Louis Pasteur develops the first artificially produced vaccine.

U.S. PHYSICIAN GEORGE STERNBERG isolates the bacteria pneumococcus, which causes pneumonia.

82

GERMAN BIOLOGIST WALTHER FLEMMING publishes *Cell Substance, Nucleus and Cell Division*, which reports his discovery of chromatin, later to be called chromosomes, and mitosis, or cell division.

CHARLES DARWIN DIES.

A PLANT CELL DIVIDING
AS ILLUSTRATED IN
SCIENTIFIC AMERICAN
IN THE NINETEENTH
CENTURY.

▼

85

GERMAN MECHANICAL ENGINEER CARL BENZ builds the first working automobile with a gasoline-burning internal-combustion engine. It has three bicyclelike wheels and runs at a speed of nine miles per hour.

LOUIS PASTEUR saves the life of a boy bitten by a rabid dog by developing a vaccine for rabies.

84

PROVING THAT THE TEACHER is not always correct, Swedish chemistry student Svante Arrhenius barely earns a passing grade when he proposes in his Ph.D. thesis that atoms carry electrical charges. Chemists of the day believe atoms are featureless, but by 1903—when the rest of the chemistry world has caught up—Arrhenius's thesis has won a Nobel Prize.

FRENCH CHEMIST Hilaire de Chardonnet invents the first synthetic fiber. It is called rayon—from the French word for ray of light—because of its shininess.

THE WORLD'S FIRST roller coaster is constructed on Coney Island.

86

THE U.S. WAR AGAINST NATIVE AMERICAN tribes comes to an end with the capture of Apache leader Geronimo.

88

GERMAN PHYSICIST HEINRICH HERTZ becomes the first person to produce and detect radio waves.

89

THE EIFFEL TOWER is completed in Paris. Designed by French engineer Alexandre-Gustave Eiffel, it stands 993 feet in height and remains the tallest structure in the world for the next four decades.

THE ROAD
TO THE FUTURE

With the merging of computers and telephones, information technology is in the process of significant change. In principle, it is now possible to have an electronic network connecting every individual and organization with a computer and a modem, transcending geographical, political, social, and temporal **Gary T. Marx** borders. Instant communication at any time, from anywhere, is possible. Databanks containing much of the world's knowledge, culture, and entertainment can be drawn upon. Work, shopping, and learning can be carried out without leaving home. Once fiber-optic cable is wired universally, it will be possible for computers to carry moving images and sound efficiently, in addition to data, the printed word, and graphics.

When computers are connected to each other they constitute a network. The Internet, for example, consists of thousands of computers globally linking millions of people. It was originally designed to tie military research sites, and later universities, into a high-speed communications network. It has since broadened to include commercial and private interests as well as individuals.

In 1993, President Bill Clinton announced plans to create the National Information Infrastructure, popularly referred to as the data superhighway. In fact, there will likely be multiple intertwining roads, with the Internet serving as the prototype. The highway will involve computer networks, cable television, interactive telephones, and still newer technologies. At this early stage in development, we can only say with certainty that the familiar communication worlds we know today will soon be drastically altered.

Currently, telephone networks permit direct interaction among dispersed individuals. Cable television offers greater diversity in program choices. Video- and audio- recording devices permit the saving of information for replay at convenient times. But the genuinely new and challenging aspects of the data superhighways are the scale, ease, and efficiency of communication, the enhanced interactivity and choice, and the breakdown of traditional concepts and borders. Over data superhighways, the volume of inexpensive, instantaneous personal communication that can be sent or received is limited only by the number of people having access

◄

IN THE LAST 150 YEARS, DATA HAS TRAVELED VIA A NUMBER OF DIFFERENT ROUTES. THE ENVELOPE *(FAR LEFT)* TRAVELED 3,400 MILES VIA BICYCLE MESSENGERS IN 1896.

to a network. Virtually any database that is part of a network, in principle, could be made available.

The data superhighway will change dramatically the basic method by which communication is exchanged. Traditional radio and television are mass media of vertical communication. They rely on a central provider, and the information flow is one-way—offering standardized material to unconnected individuals at a fixed time. With the new technology, the distinction between producers, distributors, and consumers of information breaks down. Horizontal communication among dispersed individuals avoids central control. Even traditional one-way communication such as entertainment or news can be programmed to appear at the user's discretion. In addition, the distinction between the telephone, cable television, and computer will no longer be clear. Communicating through a computer involves talking through writing; as such, it is a new form, although it is related to the telegraph. We are not certain how to think about it. Should electronic mail, e-mail, for example be viewed as a post-card, a first class letter, or a telephone conversation? Is a posting on a bulletin board best seen as a form of publishing or simply a method of conversation?

These are commonly asked questions, because technology often outpaces social customs and the law. There is a lack of agreement about how best to think about the new networking.

Will information highways be freeways or toll roads? Will the predominant form resemble a public library in which everyone will have easy access, or will it become a commodity similar to pay-per-view cable television? Will we see a new form of inequality between the information available to the rich and poor?

Who will provide the signals and services? Possible providers include telephone, cable television, or entertainment companies, non-profit organizations, government, or various combinations.

Will there be balance among the various potential uses? These uses could include interactive communication for far-flung citizens, education, public-service information and discussions, shopping, and entertainment. Will commercially driven uses predominate?

Will electronic communication be covered under the Fourth Amendment's protection against unreasonable search and seizures? Will communications over a network be given the freedom of speech and assembly protections of the First

Amendment? Can children be encouraged to use networks, yet be protected from exploitive communication?

The ease of downloading and printing out whatever one encounters on a computer screen offers new temptations for re-use and alteration not associated with printed materials or with face-to-face conversation. Should electronic communications and supporting software be entitled to the same copyright protections a book is given?

Electronic trails left behind on the data superhighway create unprecedented possibilities for knowing where a person is, with whom they are communicating, and what is being expressed, as well as what information they are accessing. Such information has enormous commercial and law-enforcement value. Will systems be technically, legally, and socially designed such that their advantages do not come at a cost to personal privacy? Can systems be user friendly and inexpensive, yet secure?

The amount of available information and the number of possible communicators available on the superhighway is staggering. Will the new "virtual" communities and interactions that occur in cyber-space mean greater equity? Race, gender, age, and physical condition are not readily apparent on a computer screen. Can we expect increased chances for social participation, and reduced social isolation? Will such interactions be as satisfying as those in the world of face-to-face interaction? Or will social skills decline and interaction become more mechanical and emotionless as a result of being electronically mediated? Will social fragmentation increase? What will be  the consequences of the blurring of traditional boundaries between work and home and the difficulty of nation states in controlling the information flow across and within their borders?

Because these developments are still in their infancy, we don't know how they will be defined, what form they will take, or what consequences they will have. Judging from earlier predictions about technologies like radio and automobiles, there will be unanticipated effects, and some current predictions will prove to be groundless. But it is clear that the nature of communication is undergoing qualitative changes that are likely to be as significant as the changes resulting from the invention of speech, writing, the printing press, and the telegraph—inventions on which data highways depend and toward which they extend.

THE LIFE AND DEATH
OF THE DINOSAURS

Richard Owen was the greatest comparative anatomist and theoretical biologist of
Victorian England. He argued staunchly against transmutation (one species giving
rise to another), even though he accepted some forms of evolutionary change.
In fact, he "invented" the Dinosauria, in part to demonstrate the error of
evolutionary progressivism—the idea that evolution necessarily means change toward higher and better forms. In 1842, just more than 150 years ago, Owen
erected the Dinosauria to accommodate the remains of three

**Kevin
Padian**

scrappy dinosaurs that had been found in England. Although
they were incomplete, Owen recognized important features about
them that separated them from all other reptiles. They were
not only very large, but also terrestrial, not aquatic. Dinosaurs also had five
vertebrae in their hips, while other known reptiles had only two. The hips and
hind legs were organized so that dinosaurs had to stand upright like mammals and
birds—they could not sprawl like other reptiles. By showing that ancient, extinct
reptiles were more "advanced" than living reptiles, Owen refuted progressivism
and sowed confusion among researchers about the process and meaning of
evolutionary change. Owen's analysis of the Dinosauria still stands today, more
than 150 years and 400 dinosaurs later.

Dinosaurs have been found from the Equator to the Poles, in every continent
and in all time periods, from the early Late Triassic (about 225 million years ago) to
today. In fact, one of the most interesting ideas about dinosaurs to emerge the past
few decades is that dinosaurs are still alive; but in fact, this is also a very old idea.

In the mid-1860s, Thomas Henry Huxley, a comparative anatomist and
paleontologist, was studying the leg bones of the meat-eating dinosaur
Megalosaurus when he noticed the close resemblance they bore to the bones of
ostriches. Huxley proposed an evolutionary link between dinosaurs and birds,
but did not have enough evidence to convince his colleagues. A century later,
John Ostrom of Yale University assembled a similar case, this time based on much
more extensive evidence. Cladistic analysis, an important new method of testing
hypotheses of evolutionary relationship, has confirmed Ostrom's view through more
than 200 unique characteristics that birds share with dinosaurs, and the definitive
placement of birds among the carnivorous dinosaurs becomes better established
as new dinosaurs and their features become better known. So dinosaurs did not

◄

become entirely extinct by the end of the Cretaceous Period; they are as familiar to us as birdbaths and Thanksgiving dinners.

Explorations in the middle of the nineteenth century brought to light the first full skeletons of dinosaurs, and confirmation that, again unlike any other reptiles, many of them stood solely on their back legs. The forelimbs, particularly of the meat-eaters, the duckbills, and their relatives, were simply too small and too unusually adapted to have supported the animals in walking and running.

As the study of dinosaurs continued, by about the turn of the twentieth century, it became clear that they had achieved great diversity over the course of 165 million years. Meat-eating theropods came in sizes ranging from the chicken-like Compsognathus to the giant Tyrannosaurus. The duckbills, sometimes called the "cows of the Cretaceous," were numerous and sharply distinguished from each other by elaborate skull crests, horns, hooves, and probably also by colors and sounds. The armored ankylosaurs, the plated stegosaurs, the dome-headed pachycephalosaurs, the horned and frilled ceratopsians and the giant herbivorous sauropods are familiar to several human generations. Elmisaurus, Troodon, Segnosaurus, Baryonyx, Saltasaurus, Carnotaurus, Jurassosaurus and a host of others have been discovered since the "renaissance" in exploration since the 1970s—at the rate of nearly one new dinosaur discovered every six weeks.

Dinosaurs were not like the reptiles we know today. They were bigger, more diverse in shape and adaptation. They walked differently, and many had a different social organization from that of typical reptiles. They were not exactly like living crocodiles, lizards, and turtles, but they also weren't all like birds, their only living members.

Some dinosaurs, such as stegosaurs and ankylosaurs, are found only rarely and as isolated skeletons. In life they may have been solitary, or have traveled in small upland groups. Other dinosaurs, such as duckbills, sauropods, and ceratopsians, seem generally to have been gregarious. Their remains are found together, piled in the dozens, hundreds, or even thousands. Many dinosaurs may have had a complex social structure for at least part of the year, which is not difficult to imagine (fishes, newts, and ants do likewise).

Some of the most interesting evidence of dinosaur behavior comes from their eggs, nests, and embryos, which are turning up all over the world. It had long been assumed that dinosaurs laid their eggs in the sand and then wandered off and forgot about them, leaving the young to fend for themselves. Some reptiles do this, but many don't; and birds and crocodiles, the two closest relatives of Mesozoic Era dinosaurs, are as conscientious in their parenting as mammals.

Studies of two Cretaceous ornithopods that shared the same nesting grounds reveal sharp differences in their parental ecology. The hypsilophodont Orodromeus laid eggs in which the embryos developed quickly and ossified their bones before hatching; these babies popped out of the egg and left the nest immediately. But the hadrosaur Maiasaura laid eggs in which the embryos developed slowly and hatched with their bones still soft and cartilaginous, in a helpless state; they were apparently fed by the parents until they grew much larger than hatchling size, rolling around the nests and trampling their eggshells to fragments. Caches of juvenile skeletons found in the nests suggest that the juveniles died of starvation when parents did not return.

The fine structure of dinosaur bone reveals many differences from the bones of typical reptiles, and interesting similarities to those of birds and mammals. The long bones are formed of a woven tissue of bone that is very well vascularized, like the bones of birds and mammals. Other reptiles have compact, tree-ring-like bone with very few blood vessels. Judging by today's standards, dinosaurs grew quickly, and probably had metabolic rates that would remind us more of active mammals and birds than of lethargic, sporadically active crocodiles and lizards. But such generalizations by themselves are surely oversimplifications. More evidence is needed for a complete picture of dinosaur biology and behavior.

If dinosaurs were so successful, why did they become extinct? At the end of the Cretaceous Period, a large asteroid, or several, almost certainly struck the Earth, causing destruction of incalculable dimensions. But this catastrophe did not substantially affect the dinosaurs, because (except for the birds) as far as our records show, they were already extinct, and in fact their diversity had been dwindling for millions of years. The last years of the Cretaceous don't show heightened extinction rates; but they do show a precipitous drop in appearances of new species, which amounts to the same effect, but through very different causes.

The asteroid that drew the Cretaceous to a close may have caused havoc in the marine realm, but the signals are less clear on land. Mammals, crocodiles, turtles, fishes, lizards, amphibians, and many otherwise sensitive indicators of environmental crisis seem to have weathered this shock, with no apparent catastrophic drop in diversity. All these groups survived and flourished into the Tertiary. Mass mortalities there may have been; but on land, the notion of a mass extinction seems overdrawn. There are no mass graveyards at the Cretaceous-Tertiary boundary: apparently the dinosaurs were already gone.

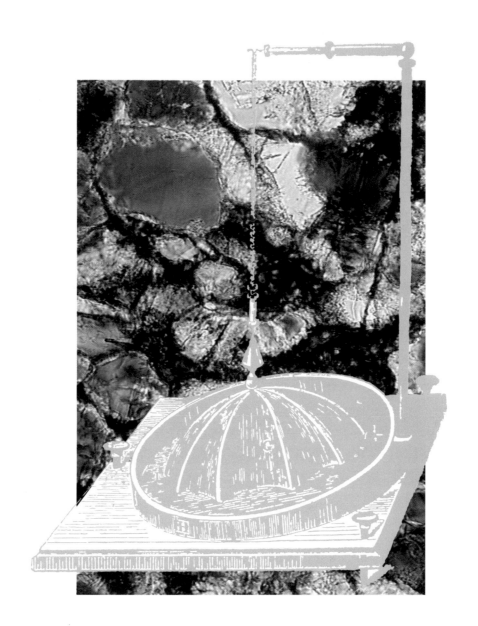

GREAT SHAKES

Over the past 150 years, earthquakes have killed more than two million people and many millions have had their local economies destroyed as a result of earthquake devastation. As world population increases, the threat of earthquakes grows. At the same time, amazing progress has been made in our understanding of the causes of earthquakes and their properties. Today we know a great deal about how to build structures that can withstand earthquakes. Earthquake research has contributed, also, to geological knowledge: analysis of seismic waves traveling through Earth have given us details about the Earth's deep interior.

Bruce A. Bolt   Seismology, the scientific study of earthquakes, took its first leap forward when a distinctive earthquake struck near Naples in 1857. Robert Mallet, an Irish engineer who spent three months in the damaged area, laid much of the basis of field seismology as we know it today. Mallet concluded, "When the observer enters upon one of those earthquake shaken towns he finds himself amidst utter confusion. Houses seem to have been precipitated to the ground in every direction. There seems to be no governing law. But it is only by patient examination house by house analyzing each detail that at length we perceive that this apparent confusion is but superficial."

In the same year, a major earthquake struck a sparsely populated region of California, north of the city of Los Angeles. A crack in the ground 70 kilometers long opened near Fort Tejon. We now know that this crack was caused by the horizontal slip of the San Andreas fault which runs from the Gulf of California in Mexico to northern California. This immense fault is the boundary between two tectonic plates of the Earth's outer shell, the Pacific plate and the North American Plate. Only after the great San Francisco earthquake in 1906 was it recognized that sudden surface fault rupture, in this case extending for over 400 kilometers, caused the ground to shake instead of being the result of the shaking.

The basic explanation for earthquakes—an explanation that dominates present seismological theory—is that the seismic waves are generated by a rapid, deep slip along the fault rupture. Before hand, surrounding rocks are elastically strained by slow tectonic deformation. After hundreds and even thousands of years, the strength of the already-fractured rock along the fault zone is not sufficient to prevent a sudden displacement. This "elastic rebound" theory of earthquake genesis explains almost all earthquakes, even those associated with volcanoes. Exceptions are

◄

A NINETEENTH-CENTURY EARTHQUAKE INDICATOR IS SHOWN OVER A FAULT ZONE CREATED BY DEFORMING A SAMPLE OF MANTLE OLIVINE.

man-made earthquakes, such as those from underground chemical or nuclear explosions, rock avalanches, and other sudden forces at the surface.

At the end of the eighteenth century, the invention of seismographs, instruments that measure earthquake waves, uncovered central insights into the attributes and causes of earthquakes. Unexplained ground shaking, which had inspired terror of the unknown, was depicted in wavy lines traced on film or tabulated as a list of digital numbers in a computer memory. At the turn of the century, the first global network of seismographic stations went into operation.

After the First World War, the worldwide monitoring system was considerably expanded. By 1920, about 80 stations operated around the world, each reporting to a central organization which identified earthquake locations by triangulating the seismic wave arrivals, not only from inhabited areas, but also from remote regions, even under oceans. A complete distribution of global seismically active zones then emerged.

In 1935, the American seismologist Charles Richter established a simple magnitude scale for identifying the strength of earthquakes. Today, several other methods, using more quantitative techniques, also define earthquake magnitude. In the first half of this century, seismic-wave research was applied to oil and mineral prospecting and to understanding the deep internal structure of the Earth. By 1936, measurement of earthquake waves had helped scientists to determine the depth of the molten central core and the existence of a solid inner core.

With the coming of the atomic age, seismographs were used to record the artificial earthquakes produced by nuclear underground tests. A complete test-ban treaty would require discriminating between seismic waves produced by natural earthquakes and those created by clandestine underground nuclear explosions.

Severe Shaking Lasted 40 Seconds

The duration of the 1906 San Francisco earthquake is known from two remarkable observations by scientists living in the area. A professor of astronomy at the University of California at Berkeley reported, "I counted 40 seconds of strong shaking as I carried my children out of the house." Soon after the earthquake, the head of the San Francisco Weather Bureau wrote, "My custom is to sleep with my watch open, notebook open at the date, and pencil ready, also a hand torch. They are laid out in regular order, torch, watch, book and pencil. I entered in the book 'severe shaking lasting 40 seconds.'"

The use of computers and improved global seismograph networks have increased the accuracy of determining earthquake sources and fault mechanisms that release seismic energy. Faults that generate earthquakes along mid-oceanic ridges, for example, have very different slip than do faults that produce earthquakes below the deep oceanic trenches. By 1970 such inferences offered scientists a solid basis for the revolution in our understanding of earth dynamics—plate tectonics. Plate tectonics now provides order for various patterns and rates of occurrence of earthquakes and volcanoes. It has allowed researchers to make predictions of the

types, location, and frequency of earthquakes and volcanic activity that might occur in a specified region. It has also offered explanations for geological questions such as how mountains are built—and it has helped to reduce risk associated with earthquake activity.

Earthquake prediction is not a new idea, but with the introduction of improved seismographic equipment, major programs aimed at forecasting were begun in many countries beginning with the 1960s. Scientists in many countries, particularly Japan, have explored a variety of possible indicators—

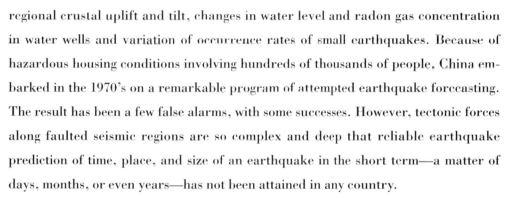

▲
THE KUTCH
EARTHQUAKE OF
1819 REDUCED
SINDREE FORT
(TOP), WHICH
STOOD ON A RISE,
TO A SUBMERGED
RUIN *(BOTTOM)*.

regional crustal uplift and tilt, changes in water level and radon gas concentration in water wells and variation of occurrence rates of small earthquakes. Because of hazardous housing conditions involving hundreds of thousands of people, China embarked in the 1970's on a remarkable program of attempted earthquake forecasting. The result has been a few false alarms, with some successes. However, tectonic forces along faulted seismic regions are so complex and deep that reliable earthquake prediction of time, place, and size of an earthquake in the short term—a matter of days, months, or even years—has not been attained in any country.

In contrast, prediction of future earthquake ground motions has helped to create building codes to make new construction safer. Still, recent earthquakes show the enormous task facing hazard reduction. In 1976, alone, earthquakes in Guatemala, Italy, and China killed more than 300,000 people. In 1985, Mexico City lost more than $4 billion, and in California, where building practices are regulated, the Northridge 1994 earthquake caused $10-15 billion damage. Yet, despite the remaining technical difficulties in predicting future earthquakes and the technical gaps in earthquake-resistant engineering, there are no scientific or engineering reasons why earthquake risks cannot be reduced significantly.

1890

**U.S. SURGEON
WILLIAM HALSTED**
introduces the practice of
wearing sterilized rubber
gloves during surgery.

93

AUSTRIAN PHYSICIANS
Josef Breuer and Sigmund
Freud publish *On the
Psychical Mechanism
of Hysterical Phenomena*.
The paper becomes the foun-
dation for psychoanalysis.

95

**EXPERIMENTING
WITH A CATHODE RAY
TUBE,** German physicist
Wilhelm Roentgen detects
invisible radiation that
passes through layers of
paper and even metal.
Unable to identify this
powerful energy, he adopts
the algebraic symbol X,
designating an unknown
quantity and refers to his
discovery as X-rays. His
finding sends the scientific
community into a frenzy,
and within days of his announcement doctors begin
using X-rays to help them
see inside the human body.

**RUSSIAN PHYSICIST
ALEXANDER
STEPANOVICH** and
Italian electrical engineer
Guglielmo Marconi develop
the antenna to send and
receive radio waves.

LOUIS PASTEUR DIES.

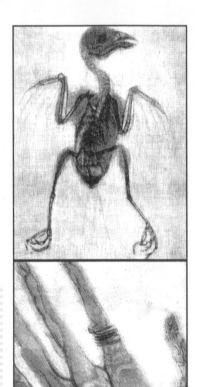

▲
X-RAYS OF A BIRD (*TOP*)
AND A HAND (*BOTTOM*),
WHICH APPEARED IN
SCIENTIFIC AMERICAN
IN 1896

96

SVANTE ARRHENIUS DISCOVERS that the amount of carbon dioxide in the atmosphere determines global temperature. He theorizes that past ice ages occurred because some event had reduced the level of carbon dioxide.

97

BRITISH PHYSICIST JOSEPH THOMSON discovers the first subatomic particle, the electron.

GERMAN CHEMIST EDUARD BUCHNER finds that a cell-free extract of yeast will convert sugar into alcohol—disproving the prevailing wisdom that vital processes can take place only inside living cells. Buchner's discovery marks the beginning of biochemistry, and from this point on, the term enzyme is used for all biochemical catalysts both inside and outside the cell.

98

FRENCH CHEMISTS MARIE AND PIERRE CURIE coin the term radioactivity and discover the radioactive elements polonium and radium.

DUTCH BOTANIST MARTINUS BEIJERINCK becomes the first scientist to identify a virus—which he names after the Latin word for poison—when he traces the infective agent that causes tobacco mosaic disease.

MARIE AND PIERRE CURIE

99

GERMAN MATHEMATICIAN David Hilbert develops the basic concepts of geometry in his publication *Foundations of Geometry*.

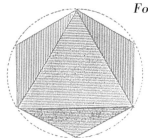

MAKING THINGS
WORK BETTER

Energy is, of course, a physical concept, but the notion that it should be conserved derives from economic considerations, practical and theoretical. The actual practice of conserving energy—using less of it to achieve the same purpose—is popular only when the cost of energy is high, as it was following the 1973 oil crisis. The practical link between energy and economics reflects a basic

**Barry
Commoner**

theoretical connection: that energy's only use is in doing work; that work is essential to production (and indeed to every other human activity), and that production is the source of economic wealth. Since energy is therefore an unavoidable cost of production, when the price rises, it makes sense to use it more efficiently—to use less energy per unit of production.

The conservation response to the more than three-fold increase in real (that is, discounted for inflation) energy prices between 1973 and 1983 seemed impressive. In the United States, the total use of residential energy per household decreased by 15 percent in that decade, largely because of measures to improve energy efficiency. For example, over that period, the efficiency of a typical gas furnace increased from about 60 to 75 percent, and the efficiency of insulation in new homes nearly doubled. There were even greater improvements in U.S. industry. Between 1973 and 1983, while industrial output increased by 16 percent, the amount of energy used to produce it fell by 25 percent—a 55 percent improvement in output per unit energy used; that is, in energy efficiency.

However, since 1983, there has been little or no further improvement in residential or industrial energy efficiency in the United States, suggesting that the limit of this useful conservation practice had been reached. Yet, in 1974, when an American Physical Society taskforce published the first comprehensive review of U.S. energy conservation, they concluded that residential energy efficiency could be improved not by 15 percent, but by more than 1,000 percent. The explanation of the huge discrepancy between what was achieved in energy conservation and what could be achieved is neatly contained in the two basic principles that frame the science of energy—thermodynamics.

The First Law of Thermodynamics asserts that energy is indestructible; no matter how it is used, the amount remains the same. Energy is always conserved. The Second Law explains why energy conservation is nevertheless important.

◄ ▲

PRACTICING ENERGY
CONSERVATION IN OUR
HOMES—LIKE THIS
COTTAGE FEATURED IN
SCIENTIFIC AMERICAN
IN 1887, BECOMES A
POPULAR PRACTICE
WHEN THE COST OF
ENERGY IS HIGH.
COAL (*ABOVE*, HERE
BEING WASHED) WAS
THE CHIEF ENERGY
SOURCE OF THE NINE-
TEENTH CENTURY.

It states that the unique property of energy is its ability to do work: to lift a weight, move an auto or warm a house on a cold day—in other words, to make things happen that do not happen by themselves. The Second Law also states that when work is done, although the energy itself is conserved, its ability to do work is not. Some of that ability is always lost.

Nearly all of the gains in energy conservation, especially in residences, have been based on only the First Law. As the Physical Society report points out, from the viewpoint of the First Law, "...the task of minimizing energy consumption appears to be primarily one of hoarding." So, for example, if the energy-requiring task is warming a room, then the First Law instructs us simply to avoid losing energy enroute from the furnace to the room, for example by insulating the heating

▲

THIS DEVICE FOR
UTILIZING WATER
POWER WAS DESIGNED
IN 1889.

duct. Since such measures can generally deliver all but 40 percent of the furnace heat to the room, the First Law efficiency is 60 percent.

However, the Second Law tells a very different story. From this viewpoint, what must be conserved is not the energy itself, but its ability to do work. Therefore, efficiency should be measured by asking how much work is used to heat the room in comparison with the minimum amount of work needed to maintain it at 70 degrees Fahrenheit—against, let us say, a 40 degrees Fahrenheit outside temperature. Calculated this way, a typical home-heating system is not 60 or 75 percent efficient, but only 7 percent efficient; the work actually used is more than 14 times the work needed. Properly computed, in keeping with the Second Law, most common uses of energy operate at efficiencies of 10 percent or less. The efficiency of heating water is typically 3 percent; of refrigeration, 4 percent; of driving an automobile, 10 percent.

The Second Law not only shows that there is a great deal of room for improvement; it also points out how that improvement can be achieved. Work is available from energy when it flows from a hot place to a cooler one—as it does spontaneously. According to the Second Law, the amount of work available from the energy depends on the difference between the two temperatures. For maximum efficiency, the amount of available work—and therefore the temperature difference—should be matched to the amount needed by the work-requiring task.

The chief reason why a furnace is very inefficient is that it operates at about 500 degrees Fahrenheit in order to warm a room to only 70 degrees Fahrenheit— a form of thermodynamic overkill that wastes most of the fuel's available work. A much more thermodynamically efficient way to heat a home is the heat pump. Like an air conditioner, which pumps heat from a relatively cool room into the warmer outside air, the heat pump moves heat from cooler outside air into the warmer room. This temperature difference is much smaller than the more than 400 degree gap between the room and the furnace. As a result, the heat pump operates with a Second Law efficiency of about 50 percent.

The main lesson that the Second Law teaches us is that in order to use energy efficiently, attention must be focused on the work-requiring task, not merely on the device chosen to accomplish it. When the task is thermodynamically defined, the device built to accomplish it can be thermodynamically efficient. In a washing machine, for example, there are two thermodynamic tasks: driving the motors that agitate the clothes and pump the water, and heating the water. The two tasks have very different thermodynamic requirements. Almost all of the energy represented by a motor's motion is in the form of work and is, so to speak, high-quality energy. It can be supplied at nearly 100 percent efficiency in the form of electricity, which is itself also a form of motion—of electrons. Thus, the use of electricity to drive the washing machine motors is thermodynamically efficient.

On the other hand, as we have seen, the hot water is produced very inefficiently by using a very hot flame (high-quality energy) to produce hot water of a much lower temperature (low-quality energy). This situation can be considerably improved by designing a system that matches the source of the needed forms of energy—electricity and hot water—to the different tasks. This can be done by producing both the electricity and the hot water using the same source of energy.

The device that can accomplish this—a cogenerator—can be made from an ordinary gasoline or diesel engine. An electric generator attached to the engine driveshaft produces electricity; the heat ordinarily dissipated by the engine's cooling system is used to produce hot water—free of charge, so to speak. The energy machine—the cogenerator—is nicely matched to the energy-requiring task: operating the washing machine. Cogeneration is a highly effective means of conserving energy; it is used in some industrial applications, but so far is found in only a few residences.

Thus, energy conservation is not simply a matter of insulating hot steam pipes or cold refrigerators. It requires the creation of systems in which the sources of energy are thermodynamically matched to the energy-requiring tasks. In such systems, vehicles, for example, would be driven by electricity generated by power plants that would deliver their discarded heat to heat-requiring tasks such as warming homes. Apart from a few isolated examples (in New York City, some electric power plants also distribute heat to residential buildings), our systems of production—the work-requiring tasks that support society—are yet to be organized according to this principle.

CONDUCT
AND CONSEQUENCES

In recent years, ethical debates that began decades ago over the conduct of research and the consequences of scientific discoveries have taken on new dimensions. The debates are ever-more intense, suggesting a growing ambivalence about the meaning of ethical responsibility, the terms of scientific accountability, and the appropriate role of science in human affairs.

In the past, much of the concern about the ethical conduct of research focused on the practice of experimenting on humans. Following the disclosures of gruesome Nazi experiments during the Second World War, the Nuremberg **Dorothy Nelkin** Code and subsequent research guidelines sought to protect research subjects by requiring voluntary informed consent, limiting research on vulnerable persons, such as children and prisoners, and establishing institutional review boards to monitor research procedures. But the meaning of "voluntary" and the definition of "informed" are still contested issues.

Is the participation of those who need money really voluntary? What is the meaning of information when there are unknown risks? How vulnerable are volunteers? As research today extends to the fetus and the embryo, the definition of "vulnerable subjects" is confounded by moral debates about the meaning of "personhood" and the definition of when life begins. Despite broad agreement about the importance of protecting the rights of human subjects, public disclosures of ethical misconduct, as in the case of the Cold War radiation tests, suggest the difficulty of controlling the actual practice of human experimentation.

Another heated issue concerns the use of animals as research subjects. The antivivisection movement formed in the late nineteenth century to oppose animal experimentation. Today, animal-rights advocates engage in vociferous protests, and sometimes threatening actions, to stop animal research. Defining animals as vulnerable subjects and sentient beings, activists question the ethics of such research. In response, scientists have limited their use of animals but defend this type of research as ethical because it reduces the need for experimenting on humans.

To many scientists, ethics means simply doing good science and conducting research with integrity. But increasingly, the definition of ethical scientific behavior has come to include considering the consequences of discovery and the potential abuses of scientific information. The responsibilities of scientists who create

◄
ADVERTISEMENTS APPEARING IN THE 1880S MAKE DUBIOUS CLAIMS OF SCIENTIFIC DISCOVERIES.

knowledge that can be used to develop destructive technologies was a troubling question following the Manhattan Project and the development of the atomic bomb. After the Second World War, scientific contributions to chemical and biological warfare continued to raise concerns about the ethics of military research.

Today, attention has turned to the potential consequences of discoveries in genetics. Genetic research enables us to predict future diseases and behaviors, but can also lead to potential discrimination on the basis of biological predisposition. Genetic therapies and gene manipulation offer hope to those with genetic diseases, but they have also evoked "Frankenstein"-type images and caused concern about a revival of eugenic policies of reproductive control. In response, the Human Genome Project supports a program to study the ethical, legal, and social implications of genetics, and to make recommendations for controlling abuses of genetic information.

Recent changes in the political and economic contexts of research are adding new dimensions to old ethical questions about both the conduct and consequences of research. These changes follow from the blurring boundary between scientific discoveries and their commercial applications. Discoveries in genetics lead quickly to tests for the detection of genetic disease well before therapies are available. Research in biotechnology employs techniques directly applicable to the development of pharmaceutical, food, or animal products, often before the health implications are fully understood. Commercial interests are driving research in rapidly developing fields, and scientists often work in collaborative arrangements with industry sponsors. Some are directly involved in the commercial exploitation of their own research. The stakes of discovery include potential financial profit as well as professional prestige.

The conflicts of interest that arise from these economic arrangements are undermining the openness and sharing of data that has allowed scientists to regulate the conduct of research and encouraged them to consider its social consequences. Commercial possibilities, coinciding with reduced government resources, have increased the intensity of competition in science. And proprietary interests are generating questions of ownership, royalties, and rights.

In this environment, incidents of scientific misconduct have proliferated, ranging from inappropriate claims about the benefits of research to outright cases of fraud. Public disclosures of misconduct in the treatment of human subjects are distressing testaments to the willingness of some researchers to operate outside ethical boundaries. And in some areas of current research, such as the testing of alternative cancer therapies, complex and large-scale collaborations are difficult to manage and control. A 1993 survey found that problems of misconduct were far more prevalent than many scientists had believed, reflecting fundamental changes in the scale, structure, and support of scientific projects, and in relationships among participating scientists. These changes are challenging longstanding assumptions about the ability of scientists to control the conduct of their research through the traditional mechanism of the peer-review system. Many researchers are calling for a reexamination of professional society guidelines and publication practices.

The growing intensity of ethical debates reflects, in part, the increased focus of research on sensitive areas of human biology. It also points up the power of the media to turn incidents of fraud and abuse into very public affairs. But ethical debates also reflect deeper public concerns about the role of science in society, the relative costs and benefits of research, and the ability of scientists to regulate themselves.

These concerns are manifested in the proliferation of citizens' groups seeking to influence the conduct and consequences of research. Creationists oppose the teaching of evolution theory in public schools, demanding "equal time" for their religious views. Anti-abortionists vehemently oppose fetal experiments and the use of embryos in research. Advocacy groups representing those suffering from specific diseases demand increased funding for research on their problems. Gay-rights activists challenge the procedures underlying research on AIDS therapies. Associations championing the rights of the disabled question the discriminatory consequences of discoveries in genetics. Religious groups oppose the applications of research in biotechnology that involve "tampering" with life.

▲

ETHICAL ISSUES
ARISING IN SCIENCE
AND TECHNOLOGY
TODAY MAY BE
DEBATED IN COURT-
ROOMS TOMORROW.

Today, the ethics of discovery is a critical issue on the science-policy agenda. Scientific agencies and professional societies have created review boards, commissions and study groups to monitor the ethical conduct of research and to recommend ways to control abuses of scientific information. And scientists, working in a context of public scrutiny, are facing demands for accountability arising from a growing public awareness of the social and ethical implications of scientific discoveries.

DARWIN'S LEGACY

Evolution is the idea that all species of plants, animals, fungi and micro-organisms are descended from a single common ancestor through a natural process of ancestry and descent. We now know that the Earth is just over 4.5 billion years old. The oldest sedimentary rocks are approximately 3.7 billion years old—and they contain fossils of bacteria. Life is an integral part of the history of the planet.

Evolutionary biology owes its own life primarily to one person, the Englishman Charles Robert Darwin. His *On the Origin of Species* (1859) convinced his contemporaries that life has, in fact, evolved. Darwin provided evidence from the comparative anatomy, embryology and geographic and geological distributions of organisms, all of which attest to the reality of evolution. The pattern of similarities that links all living species, Darwin argued, is the simple, necessary result of what he called "descent with modification"—which soon came to be known simply as "evolution."

Niles Eldredge

Crucially, Darwin was able to describe the mechanism by which plants and animals might change over time to fit their environment: natural selection. Darwin knew that variation within natural populations is largely heritable—although the principles of genetics had still not been grasped. He also understood (from the late eighteenth century writings of Reverend Thomas Malthus) that more offspring are born in each generation than can possibly survive and reproduce—otherwise the world would long ago have been overrun by members of a single species. Thus only those best suited to living in their habitats will thrive, reproduce and pass their success-giving traits on to their offspring. If the environment should change, other characteristics will then be superior, and so the traits of organisms will change as time goes by. This is the essence of natural selection—a process simultaneously discovered by Darwin's contemporary, the naturalist Alfred Russell Wallace.

The science of genetics began in 1900 with the rediscovery of the early experiments of the Russian biologist Gregor Mendel. For several decades, the early findings of genetics seemed to conflict with the Darwinian notion of natural selection. For example, early data suggested that mutations were invariably harmful and large-scale in effect—thus hardly the mechanism of new, useful variation by which selection might work. But, largely through the efforts of the mathematically-gifted geneticists Sewall Wright (USA), Ronald Fisher and

◀

CHARLES DARWIN
AND HIS FINCHES

85

J.B.S. Haldane (both UK), conflicts between the findings of genetics and the original Darwinian notion of natural selection were resolved, giving rise to "neodarwinism" in the early 1930s.

Shortly thereafter, three New Yorkers—the geneticist Theodosius Dobzhansky, the ornithologist Ernst Mayr, and the paleontologist George Gaylord Simpson—demonstrated respectively that all data from experimental and field genetics, systematics and the fossil record are consistent with neodarwinian principles. The "Modern Synthesis" reflected the consensus among the world's biologists that Darwin's formulation of evolution through natural selection had indeed been substantially correct. By 1959, the centennial of the publication of Darwin's *Origin of Species*, many evolutionary biologists openly claimed that evolutionary theory as it stood was virtually complete.

▲

But evolutionary biology has been far from static in recent years. Evolutionists, such as George Williams (US), John Maynard Smith and Richard Dawkins (both UK), have extended the neodarwinian paradigm in distinct directions. In their work, natural selection has come to be seen strictly as the outcome of competition for reproductive success. Dawkins sees such reproductive competition as occurring not so much among organisms, but actually among the genes themselves. Evolution, in the view of these biologists, is a race to leave as many copies of genetic information as possible to the next generation.

Sociobiology is a direct outgrowth of such gene-centered evolutionary theory. Social behavior had posed a challenge to evolutionary biologists all the way back to Darwin: how could the apparently altruistic—unselfish—behavior of individuals cooperating in societies be reconciled with the competitive and therefore essentially selfish process of natural selection? Biologist William Hamilton (UK) pointed out that, under some circumstances, it could benefit an organism's goals to transmit as many copies of its genes as possible if, instead of itself reproducing, the organism acted to foster the reproductive activities of its closest relatives. This notion of "kin selection" helps explain the evolution and internal structure, especially of insect social systems, where all members of, for example, a bee hive share some degree of genetic relatedness.

Modern neodarwinians have been stymied by another paradox—the very existence of sexual reproduction. If the essence of evolution is to maximize an individual organism's genetic contribution to the next generation, why do males and females mix their genes in equal proportions when they reproduce? Thus far, the gene-centered school of evolution has failed to come up with a convincing answer.

Recently, certain paleontologists, among them the Americans Niles Eldredge, Stephen Jay Gould and Elisabeth Vrba (transplanted to Yale University from South Africa), have demonstrated that evolutionary patterns in the fossil record have much to reveal about the nature of the evolutionary process. Species tend to remain quite stable, virtually evolutionarily static, over periods of millions of years. Adaptive change through natural selection does not accumulate by slow degrees as Darwin originally thought. Rather, evolution seems to be concentrated mostly in relatively brief spurts associated with the division of an ancestral species into two or more descendants: the process of "speciation" first discussed in modern terms by Dobzhansky and Mayr. Eldredge and Gould called this pattern of evolutionary stasis interrupted by brief episodes of speciation "punctuated equilibria."

Punctuated equilibria has many implications for our understanding of large-scale evolutionary phenomena ("macroevolution"). What, for example, causes long-term evolutionary trends—such as the increase in brain size in human evolution? Brain size has remained stable within ancestral human species, increasing only when new species evolved sporadically during the past four million years. Evidently, bigger-brained species had some advantage over their less-well-endowed kin, suggesting that a form of "species selection" roughly analogous to natural selection also takes place in evolutionary history.

In yet another related field, molecular biology promises to unlock still more mysteries of evolutionary history—and perhaps to reveal further secrets of the very nature of the evolutionary process itself. Modern evolutionary biology remains a lively branch of science, bursting with new ideas—and new controversies.

Too Many Elephants

How did Darwin come to know that not all organisms born can possibly survive and reproduce? Simple mathematics. Darwin supposed that if you start with a single pair of elephants, and if you assume that elephants produce an average of six offspring during a 60-year period of fertility, in the brief span of 500 years there would be 15 million elephants descended from that single original pair! Something— lack of food, disease, climatic factors— must be at work, keeping each population in check.

The human population explosion— from one million to 5.7 billion in a scant 10,000 years—reflects the simple fact that humans abandoned local ecosystems with the invention of agriculture 10,000 years ago. Human population will stabilize once again— when we have saturated the global biosphere on which our species still ultimately depends.

01

1900

GUGLIELMO MARCONI INVENTS THE RADIO when he sends radio waves from England to Newfoundland using balloons to loft his antennas into the sky.

02

GERMAN PHYSICIST MAX PLANCK LAYS the foundation for modern quantum physics when he states that electromagnetic energy radiates not in continuous waves, as had long been assumed, but in discrete, indivisible packets he calls quanta.

PIERRE CURIE measures a large amount of heat given off by radium as it emits radiation. Scientists refer to this newly discovered energy source as atomic energy.

RUSSIAN PHYSIOLO- GIST IVAN PAVLOV formulates his law of conditioned response. He demonstrates that a dog given food at the same time a bell is rung will soon be conditioned to salivate at the sound of the bell alone.

AUSTRIAN PHYSICIAN Karl Landsteiner shows that there are four types of human blood: O, A, B, and AB. Before his crucial discovery, blood transfusions often result in death and are banned in many countries. Afterward, transfusions soon become a fundamental medical tool.

THE NOBEL PRIZES ARE AWARDED for the first time. At his death five years earlier, Alfred Nobel had willed virtually his entire estate of nearly $10 million for the establishment of these prestigious annual honors.

03

U.S. AERONAUTICAL ENGINEER Orville Wright makes the first controlled, powered flight in an airplane at Kitty Hawk, N.C. On his most successful trial he remains in the air for 59 seconds and covers 852 feet. The flight is made possible by Orville's work with his brother, fellow engineer Wilbur Wright, in designing engines of unprecedented lightness.

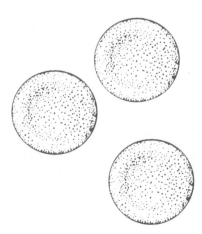

05

GERMAN-BORN SWISS PHYSICIST Albert Einstein formulates his theory of special relativity, which states that the speed of light is constant under all conditions and time must pass at different rates for objects in constant relative motion. Einstein consequently theorizes that mass must be viewed as a highly concentrated form of energy, and he establishes the celebrated formula $E=mc^2$—energy equals mass times the speed of light squared.

07

U.S. CHEMIST BERTRAM BOLTWOOD establishes radioactive dating as a means for discerning the age of rocks and archaeological remains.

08

U.S. INDUSTRIALIST HENRY FORD makes the automobile affordable to millions by developing the assembly line, which allows him to mass-produce cars at minimal cost. His first product, the Model T, is initially priced at $950 but drops to $290 after only a few years. The automobile age has begun.

09

GERMAN BACTERIOLOGIST PAUL EHRLICH attempts to cure sleeping sickness by synthesizing the compound arsphenamine. While the compound fails, Ehrlich discovers that it cures syphilis instead. Arsphenamine soon cuts in half the syphilis rate in England and France, but some decry the drug, claiming that it encourages immorality.

U.S. CHEMIST PHOEBUS LEVENE extracts a sugar from a nucleic acid and identifies it as ribose. Not all nucleic acids possess it, but those that do come to be known as ribose nucleic acid, or RNA.

U.S. EXPLORER ROBERT PEARY, his associate Matthew Henson, and four Eskimos become the first humans to reach the North Pole.

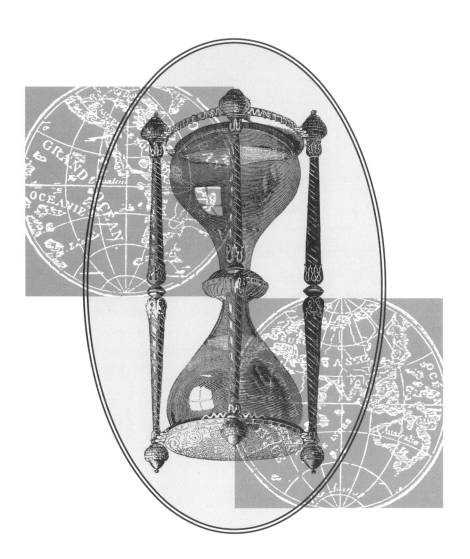

A TERRIBLE TOLL

By consuming directly, diverting, or wasting an estimated 40 percent of total net terrestrial photosynthetic productivity, human beings are driving the earth into an era of instability and precipitating the sixth major extinction event to take place in

Peter H. Raven

the 1.5 billion-year history of eukaryotic life on this planet. All of the extinction events that we can recognize in the fossil record took place after the evolution of multicellular life. The first occurred near the end of the Cambrian Period (about 505 million years ago) and the second marked the close of the Ordovician Period (about 438 million years ago)—times when all multicellular life existed in the sea. The third major extinction event came at the end of the Devonian Period, about 360 million years ago.

The most drastic of all extinction events that we have detected happened during the last 10 million years of the Permian Period, which ended the Paleozoic Era, about 248 to 238 million years ago. Paleontologists estimate that as many as 96 percent of all species of marine organisms became extinct at that time, permanently altering the character of life on earth.

Sixty-five million years ago, at the end of the Cretaceous Period, which also marked the close of the Mesozoic Era, a major extinction event took place. It was probably linked with the impact of a huge asteroid, possibly near what is now the Yucután Peninsula, and it eliminated a high proportion of all marine organisms and perhaps two-thirds of all the species living on land at that time. Dinosaurs ceased to be, and mammals, none of which was larger than a house cat at the time, rose to a dominant position on land, while birds, modern reptiles, plants, and insects proliferated greatly.

For the past 65 million years, the number of species of plants, animals, fungi, and microorganisms on earth has been increasing steadily, as has the complexity of terrestrial biological communities. During the time of the human being, genus Homo, about two million years of the earth's four and a half billion-year history, there have probably been about 10 million species of eukaryotic organisms in existence, perhaps 15 percent marine and the rest terrestrial or freshwater. Never in the history of life have more species occurred at any one time.

When our ancestors developed agriculture about 10,000 years ago at several widely scattered centers, it is estimated that the total human population consisted of perhaps several million people—equivalent to that of a medium-sized city of the twentieth century. When Stonehenge and later the Pyramids were being constructed, the entire population of the earth may have approximated that of present-day New York State or Taiwan—somewhat more than 20 million. At the time of Christ, our numbers had swollen to perhaps 130 million, the population of the United States 50 years ago (half what it is today), and by 1950, there were 2.5 billion of us—like all of the earlier totals, an unprecedented number.

Biological extinction—the permanent disappearance of species—has always been a feature of life on this planet, and the estimated 10 million species living today clearly constitute only a small fraction, perhaps even less than one percent, of all of the species that ever existed. However, in an extinction event, or extinction spasm, like the one we are presently experiencing, we confront extinction rates that are many times greater than usual. Why is this so?

Human pressures on the global environment are essentially the product of population level, the consumption rate per person, and the technology employed. Although only about a fifth of the approximately 5.6 billion people on earth live in industrialized countries like the United States, Japan, or the countries of Europe, these people consume, directly or indirectly, approximately four-fifths of the world's productivity; control 85 percent of the world's financial resources; and have a disproportionately large effect on the sustainability of the biosphere. While the global population has been more than doubling from its 1950 level of 2.5 billion to its present 5.6 billion people, a fifth of the world's topsoil has been lost; some 15 percent of the world's agricultural land has become so waterlogged, saline, or desertified that it is no longer arable; carbon dioxide in the atmosphere has increased by a fifth; the stratospheric ozone layer has been depleted by six to eight percent; and about a third of the world's forests have been cut down without being replaced.

Unprecedented numbers of human beings are putting a strain on the earth's capacity to sustain life that far exceeds anything that has occurred in the past, a situation of which the ultimate, and even intermediate consequences are far from certain. In the mid-1990s, about 1.5 billion people were living in absolute poverty, approximately half of them were malnourished. Meanwhile, nearly 100 million people per year are being added to the world population, which, because there are so many young people in developing countries, will almost certainly not stabilize before it reaches a level of somewhere between 8.5 and 14 billion, or even more.

The reduction of habitat caused by a human population of this size, with these characteristics, is estimated to be threatening the extinction over the next several decades of as many as a fifth of all of the species of plants, animals, fungi, and microorganisms on earth. By the end of the next century, when the human population may have stabilized, if family planning remains a top priority world-wide, we are likely to have lost even more than the two-thirds of the total species we believe became extinct at the end of the Cretaceous Period. The earth as a system is capable of capturing energy from the sun and functioning sustainably because of the properties of communities and ecosystems, which in turn are based on biodiversity and vary greatly in species composition from place to place. Humans obtain their food, medicines, building materials, clothing, and energy from myriad individual kinds of organisms, but we understand them so little that we have so far identified and named only about 15 percent of the total—much less examined their economic application or importance to us.

The present extinction rate which is closely associated with habitat destruc-tion, is estimated to be running at some 1,000 to 10,000 times the background rate (the rate characteristic of the past 65 million years), making the extinction of biodiversity the fastest-moving and most threatening of all the environmental changes now confronting us. As this extinction is completely irreversible and will critically limit human ability to organize sustainable, productive systems in the future, its progress is a trend we should be viewing with great alarm, and concerning which we should be taking urgent action.

About 80 percent of terrestrial biodiversity occurs in developing countries, which have a rapidly increasing proportion of 80 percent of the world's people, but only 15 percent of the wealth and six percent of the world's scientists and engineers. Therefore the highest priority must be placed on international cooperation if the loss of species is to be arrested. Biodiversity cannot be preserved adequately in a world that is getting warmer, losing its topsoil at the rate of 25 billion tonnes per year, or depleting its protective ozone shield; in other words, the earth as a whole must be managed sustainably in order to provide a context for the preservation of biodiversity. If that basic goal was accomplished, then the cultivation of nature reserves selected for their biodiversity, combined with the careful management of degraded or disturbed lands, and with comprehensive programs of ex situ conservation, would offer the greatest hope of saving biodiversity in the midst of the great extinction spasm that we have brought upon ourselves.

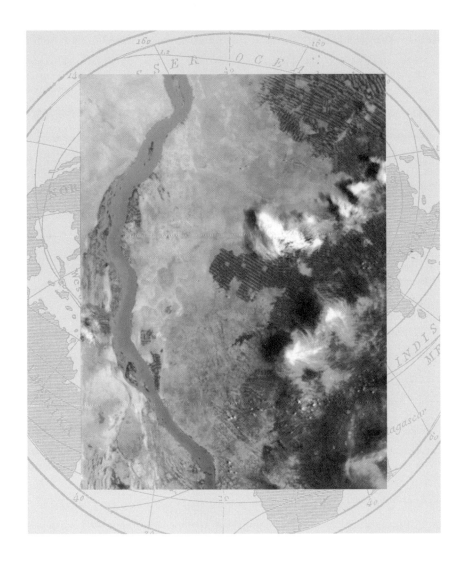

FEEDING THE WORLD

Nothing has a more extensive and far-reaching impact on our planet than producing food. To put our need for sustenance in perspective, half of the world's population are farmers and we now produce more food in a single year than was produced in the entire century before Columbus set off on his voyage of discovery.

Garrison Wilkes

With the dramatic increase in our numbers in the last hundred and fifty years have come potentially devastating environmental sacrifices to produce the food we need. Food is for us, along with clean air and clean water, a resource that sustains life. Food supplies the fuel (carbohydrates, lipids and proteins) and building materials (proteins, minerals and vitamins) that are absolutely essential.

Control over food production is a comparatively recent event—dating back only ten millennia to the agricultural revolution of the neolithic age. (Before that time, humans like all other animals, secured food by hunting and gathering.) At that point, about 10,000 years ago, we began independently to cultivate in different parts of the world a diversity of plants that we subsequently turned into crops—maize, beans, and squash in Mexico; wheat, barley, and peas in the Near East; rice, millets, and soybeans in China.

These early domestic crops were not much more productive than their wild progenitors, but the mere act of cultivating them was a radical break with the past. By now we have used in various parts of the world 5,000 plants as basic calorie sources, and of these about 500 are cultivated on significant acreage. Today, the entire world is fed by domesticated plants that 400 human generations ago did not exist except in the wild. These plants have been genetically adapted to the way we prepare the seed bed, add fertilizer and water, eradicate competing weeds and insects and ultimately collect the seed to store and plant again in the proper season.

◄

THE NILE IS THE SOURCE FOR NEWLY IRRIGATED CROPLANDS IN THE SUDAN (SHOWN IN RED IN THE SATELLITE PHOTOGRAPH).

Surprisingly, there is no one perfect food or single best diet. Depending on where we live our diet might be rice, stir-fried vegetables and tofu; bread, vegetables and milk; or tortillas, beans and salsa. But what defines our food needs is quite straightforward: between 1500 and 3500 calories of energy, approximately 60 grams of quality protein (including the 9 essential amino acids) and an adequate balance of vitamins and minerals. There is also the daily need for two liters of water, or one liter per 1000 calories consumed. This standard is simple to state but difficult to achieve; with a global population of five and a half billion individuals, one half billion fall short and experience protein/calorie malnutrition.

Two new sciences of this century have had a major impact on the food supply: genetics, which has allowed us to breed plants and animals with unprecedented yields and efficiencies; and nutrition, which has allowed us to understand how foods work in the human body and thus to define what is essential. Nutritionally, the early part of this century witnessed the discovery of the role of vitamins and minerals and mid-century the identification of kwashiorkor as a protein deficiency distinct from marasmus, the disease of starvation. Both have been found to be related to protein-energy malnutrition, or PEM, the world's most widespread nutritional problem. Deficiencies of vitamin A and iron were also determined to take their toll as diseases of malnutrition. Genetically, the most dramatic changes in food production have been the advances of hybrid corn in the U.S. in the 1930's and the so-called Green Revolution introduction of new varieties of wheat and rice in the developing nations in the 1960's.

The Green Revolution has resulted in the most dramatic increase in food production in human history, with over two billion humans affected and the increased yields alone enough to feed 800 million. Although it began in the 1960s, the revolution gained extensive publicity in 1970 when the Nobel Peace Prize was

awarded to U.S. plant pathologist and geneticist Norman E. Borlaug, whose strategy of breeding high-yielding dwarf varieties for developing nations proved an enormous success. Borlaug and other plant geneticists had found that traditional land races of wheat and rice do not make effective use of fertilizer because it causes them to lodge, or grow too tall and topple over. By incorporating dwarfing genes that gave the plants short stiff straw and enabled them to respond to fertilizer without lodging, Borlaug was able to produce higher grain yields. Other genetic improvements of the era included building plant resistance to insects and fungi, insensitivity to day length and rapid maturation—so that less water was needed and a second or double cropping could be planted. Taken as a whole, the genetic advances of the Green Revolution have helped developing countries feed their expanding populations, but they are still not the ultimate solution to feeding a hungry world.

Not only is the number of plants that feed the world becoming smaller, so is the diversity and genetic variation that allow us to breed new varieties. As we add 100 million people to the planet each year, food security has become the most urgent need in this decade. For the developed nations, population increases can be accommodated by eating lower on the food chain—which means less meat—and by consuming grains directly, but the developing world already does so. Because we have fully exploited the yield increases from fertilizer, irrigation and plant growth, the new frontier is to genetically manipulate plant physiology and roots and to replace varietal uniformity with hetero- geneity. As we approach the next century, therefore, the challenge will be to breed better varieties, encourage genetic diversity within and between crops and attempt to erase the negative environmental effects of plant mono- culture and urban intensification. Our goal should be fair and equitable food production that does not come at the expense of any one geographic region or human generation, nor threaten the planet's precious remaining uncultivated wilds. It is an awesome challenge we can and must meet.

And Twenty-Five to Grow On

The actual number of plants that feed the human population is amazingly small. Just over two dozen food plants account for 75 percent of all the plant calories we consume and 90 percent of the arable land we cultivate. They in- clude six grasses: rice, wheat, corn, bar- ley, oats and sorghum; four legumes: soybeans, peas, common beans and peanuts; two sugar sources: sugar cane and sugar beets; two tropical tree crops: coconuts and bananas; four starch roots: potatos, sweet potatos, yams and cassava; five fruits: tomatos, grapes, apples, oranges and mangoes; and two vegetables: cabbages and onions. These 25 cultivated food plants literally stand between subsistence and starvation for the human population. And because domestication has made these plants human captives that no longer exist in the wild, the exploding human popula- tion must cultivate them to assure their continuing high yield...and our very survival. They need us as much as we need them.

FUTURE POWER

Well before the middle of the next century, the world faces an energy deficit of extra-ordinary proportions. By the year 2040, the total population on earth is expected to double to about 10 billion people. With the continued industrialization of Asia, Africa, and the Americas, world energy consumption is projected to triple—to

Ronald C. Davidson

30 trillion watts—over the same period of time. This rate of consumption is expected to occur, even if one assumes large gains in energy efficiency. At our present rate of consumption, the world's known oil supply will be depleted in about 60 years, and the supply of natural gas in about 100 years. While coal reserves could sustain some of the world appetite for energy for several centuries, the problems associated with mining and the environmental pollution produced by coal-fired power plants would only aggravate an already precarious ecological balance.

Energy is fundamental to an acceptable quality of life and sustained economic development, not only for advanced industrial nations, but for the developing world as well. By any measure, we must find new sources for energy in the coming decades—sources that will augment even the maximum increase in reliance on solar and renewable energies and nuclear fission.

The search for alternative energy sources has led to a highly successful international research effort to develop fusion—the process of combining hydrogen nuclei that powers the sun and the stars—as a practical energy source. Positively charged deuterium and tritium nucleii, in a high-temperature "plasma" confined by a strong magnetic field, undergo fusion reactions. This phenomenon produces both inert helium nucleii, called alpha particles, that are trapped in the plasma, and energetic neutrons that carry heat away from the reaction. Unlike fission, no long-lived radioactive byproducts are produced in the fusion reaction. The neutrons produced in the deuterium-tritium fusion reaction induce short-lived radioactivity in the vacuum vessel and the mechanical structure surrounding the plasma. But the use of reduced-activation materials would minimize the radioactivity and would allow surface burial of the structure as low-level waste. Fusion offers the prospect of an environmentally attractive and secure long-term energy source, with a virtually unlimited fuel supply in the form of deuterium obtained from ordinary water.

International fusion research for energy applications began in earnest in the early 1950s, and progress in the use of magnetic fields to confine high-temperature

◀

OUR QUEST FOR NEW ENERGY SOURCES HAS BEEN CHRONI-CLED IN *SCIENTIFIC AMERICAN* AS DEMON-STRATED BY CAPTAIN JOHN ERICSSON'S SOLAR ENGINE *(TOP)* AND E. P. WILLIS'S PERPETUAL MOTION MACHINE *(BOTTOM)*.

laboratory plasma was greatly aided by the invention of the Tokamak magnetic confinement geometry by the Russian scientists Igor Tamm and Andrei Sakharov. ("Tokamak" is a Russian acronym for "toroidal magnetic chamber.") In 1969, researchers using the Russian T-3 device achieved plasma temperatures of 10 million degrees Kelvin and good plasma confinement. As a result of these remarkable achievements, fusion researchers in the U.S., Europe, and Japan quickly adopted the donut-shaped Tokamak geometry.

Since construction of the largest U.S. Tokamak, the Tokamak Fusion Test Reactor (TFTR), was authorized at the Princeton Plasma Physics Laboratory in the mid-1970s by the U.S. Department of Energy, the fusion power in the alpha particles and neutrons produced in laboratory experiments has increased by a factor of more than one hundred million. This rate of increase is even greater than the increase, during the same period, in computer memory density, that oft-cited benchmark of technical progress. Scientific breakthroughs and world records in plasma performance have see-sawed among the large Tokamaks in Europe, Japan, and the U.S. The U.S. moved into the lead in December 1993, when the historic experiments on the TFTR Tokamak generated 6.2 million watts of fusion power. With these experiments—which used a 50-50 deuterium-tritium fuel mixture for the first time—the U.S. surpassed the 1991 European Tokamak record of 1.7 million watts. In November, 1994, TFTR produced 10.7 million watts of fusion power, surpassing its own record set just eleven months earlier. With each of these developments in fusion power, technical issues pertaining to the scientific feasibility of fusion are being further resolved.

The TFTR device includes high-power neutral beam injectors around the periphery of the Tokamak, which are used to heat and fuel the plasma to ion temperatures approaching 400 million degrees Kelvin, and which create ion densities of 6 x 1013 particles per cubic centimeter. Once the high-energy neutral tritium and deuterium atoms enter the plasma chamber, they are ionized and collisionally heat the background plasma. The high-temperature plasma in TFTR is confined and insulated from contact with the chamber wall by a strong toroidal magnetic field with a strength of 50 kilogauss (about 100 thousand times the strength of the earth's magnetic field), which is sustained for about two seconds by external power supplies. Properties of the plasma are measured by an impressive array of noninvasive diagnostic instruments (more than 50 in all), ranging from detectors that sense microwave and light emission from the plasma, to detectors that measure energetic ions escaping from the plasma, as well as neutrons produced in the deuterium-tritium fusion process.

In these high-power experiments with deuterium-tritium plasmas there are pre-liminary indications that plasma electrons are heated collisionally by the energetic alpha particles (helium nuclei) created in the fusion process. In particular, the electrons experience a temperature increase, caused by alpha heating, of about eight million degrees Kelvin above the baseline electron temperature of 100 million degrees Kelvin obtained in otherwise similar deuterium plasmas. Heating of the background plasma by the alpha particles will be very important in future-generation power reactors and experimental facilities such as the International Thermonuclear Experimental Reactor (ITER), which is presently being designed. The ITER device will be a Tokamak much larger in size than TFTR, and it is designed to produce more than 1,000 million watts of fusion power, about 100 times larger than TFTR's ultimate design capability of 10 million watts. In ITER, the energy content of the alpha particles will be so intense that the alpha particles will heat and sustain the plasma, allowing the auxiliary heating sources to be turned off once the reaction is initiated. Such a state of "sustained burn" of a deuterium-tritium plasma is referred to as "ignition."

Several scientific and technological challenges remain in the development of fusion as a practical energy source. These challenges include the demonstration of plasma ignition and controlled burn at high-fusion power; the development of durable, reduced-activation materials for the reactor-vessel wall that will with-stand large heat and particle fluxes for several-year periods; and the demonstration of continuous Tokamak operation at high-plasma pressure. The technical implica-tions are well understood and the remaining national and international test facilities required for fusion energy development are now in the design and planning phase. The scientific feasibility of fusion is being established on TFTR and the world's other large Tokamaks, the Joint European Torus in Europe and the JT-60 Upgrade in Japan. What is required now is the next generation of facilities that will demonstrate the engineering and economic feasibility of fusion as a practical energy source.

AGE OF RELATIVITY

1911

BRITISH PHYSICIST ERNEST RUTHERFORD presents his theory that the atom consists of a positively charged nucleus surrounded by negatively charged electrons.

NORWEGIAN EXPLORER ROALD AMUNDSEN becomes first human to reach the South Pole.

13

FRENCH PHYSICIST CHARLES FABRY demonstrates the presence of a layer of ozone in the earth's atmosphere. The layer helps block the sun's harmful ultraviolet rays from reaching the earth.

14

A SERBIAN TERROR-IST ASSASSINATES Archduke Francis Ferdinand of Austria-Hungary, setting off a rapid chain of events that leads to World War I.

15

WHILE SCHOOL CHILDREN have long noticed from maps that the jigsaw puzzle pieces of South America and Africa fit neatly together, German geologist Alfred Wegener becomes the first scientist to argue seriously that the continents were once joined. His theory of continental drift is ridiculed by other geologists because there is as yet no known mechanism to explain how the earth's land masses can be displaced.

▲

AN EARLY LECTURE ON
GRAVITATIONAL LAWS.

18

**GERMANY SURREN-
DERS**, ending World War I.
Some 10 million soldiers
and civilians have died in
the conflict, and 20 million
more have been wounded.

**BRITISH CHEMIST
FRANCIS ASTON**
develops the mass spectrom-
eter, which allows him to
discover that there are
different forms, or isotopes,
of the same element.

16

**ALBERT EINSTEIN
INTRODUCES** his general
theory of relativity, which
states that the laws of
physics remain unchanged
in systems moving relative
to each other at any velocity. By including the large
scale effects of gravitation,
relativity theory now
makes it possible to understand the universe as a
whole, beginning the field
of modern cosmology.

17

**DUTCH ASTRONOMER
WILLEM DE SITTER**
shows that Einstein's
general theory of relativity
implies that the universe
must be expanding.

THE UNITED STATES
enters World War I by
declaring war on Germany.

**THE OCTOBER
REVOLUTION** transforms
Russia as Bolshevik insur-
gents, led by Vladimir
Lenin, overthrow the
government. Five years
later the Union of Soviet
Socialist Republics
(U.S.S.R.) is established.

19

**ERNEST RUTHERFORD
DISCOVERS** positively
charged atomic particles
that he calls protons.

ERNEST RUTHERFORD

CHANGING GENES
IN HUMANS

Physicians have just embarked on manipulating the human genome in order to treat disease, and on May 22, 1989, a gene was successfully introduced into a human being for the first time. The first attempt to correct a hereditary enzyme deficiency using gene modification occurred in September 1990 and the first effort to improve cancer treatment through gene modification of immune human lymphocytes happened in January 1991. These initial clinical steps and many that followed were the beneficiaries of rapid developments in molecular

Steven A. Rosenberg

biology and recombinant DNA technology that made it possible to identify, isolate, sequence, modify and insert genes into human cells. This approach to the treatment of human disease is referred to as gene therapy—a therapeutic technique in which genes are inserted into cells either to correct an inborn genetic error or to provide a new function to the cell. Enthusiastic advocates of its development predict that it will profoundly change the practice of medicine in the twenty-first century.

The application of gene therapy depends on successfully introducing a gene into cells and achieving or regulating the expression of the gene once it is inserted. Because no techniques are available yet that can uniformly deliver genes to specific cells or tissues while they are in the body, current gene therapies involve either the introduction of genes into a small area by direct injection or, more often, the insertion of genes into cells that have been removed from the body, manipulated in the laboratory and then reinjected.

Most techniques for inserting genes into cells are inefficient. They involve the trapping of genes into precipitated salts or lipids which can facilitate their uptake by cells, or the breaching of cell membranes with an electric current, which allows the genes to flow into the cell. These techniques successfully introduce genes into approximately one in one hundred thousand cells, and while they are valuable for laboratory studies, they are too inefficient for clinical application. The most effective method for introducing genes into cells involves the use of viruses which, as part of their natural life cycle, enter genetic material into cells and, by the expression of their genes, can affect cell function as well as result in the replication of new virus.

The viruses most commonly used in human gene therapy are RNA viruses (or retroviruses) which can be genetically engineered so that they retain their

◄

SCIENTISTS ONE DAY HOPE TO CONTROL HISTONE PROTEINS (SHOWN IN BLUE), WHICH CAN BOTH REPRESS AND FACILITATE ACTIVATION OF MANY GENES.

ability to insert their genetic material but lose their ability to form new virus. Depending on the retrovirus and the particular type of host cell, it is often possible to achieve efficiencies of gene introduction into cells ranging from 30 to 100 percent. While retroviruses can insert their genetic material permanently into the host genome so that the new material replicates normally along with the host genes, the infected cell must be actively dividing for gene integration to be possible. Since many human cells, such as muscle and brain cells, do not normally divide, the use of

▶

CAREFUL MANIPULA-
TION OF CHROMOSOME
POLYPLOIDY HAS RE-
SULTED IN DELECTABLE,
LARGE STRAWBERRIES.

retroviruses is limited. Other viruses, such as adenoviruses and pox viruses, have the advantage of a wider host range of infection that does not require growth of the host cell; their drawback is that only transient expression of the inserted gene occurs.

To produce the desired therapeutic effect, the inserted gene must produce a protein. Understanding how natural and artificially inserted genes are regulated and expressed is one of the central problems in modern biology. All cells of an individual (with the exception of germ cells) contain exactly the same genetic complement. Different on and off switching mechanisms of individual genes cause these genes to function differently in different tissues. The way gene function is regulated is extraordinarily complex. It involves small gene segments called "enhancers" or "promoters" that regulate the expression of other genes and they also must be inserted. In some gene-therapy applications, it is important to express genes continuously, while in others, the gene must be regulated and expressed only under certain conditions. For example, the insulin gene which is inserted into diabetics must be activated only when there are high levels of blood glucose. Researchers continue to refine their ability to engineer viruses and control gene expression, and as the technology improves, opportunities for gene therapy increase.

Because all life processes are ultimately controlled by genes, potential applications of genetic modifications to prevent or treat disease are extraordinarily varied, and individual applications of gene therapy vary depending on the unique features of the disease being treated. Transferred genes may be required only in specific tissues lacking a necessary protein, such as the lymphocytes of children born with severe combined immunodeficiency disease which lack the enzyme adenosine deaminase, or the lungs of patients with cystic fibrosis, which are missing a protein involved in ion transport. Gene insertion may be used to confer resistance to harmful stimuli; certain genes, for example, will protect lymphocytes from destructive infection by the HIV virus, while others, inserted into bone marrow cells, will protect them from harmful effects of chemotherapy. Gene insertion may also provide new surface molecules to cells to increase their ability to immunize (as would be beneficial in cancer cells) or their responsiveness to growth factors (which would be desirable in cells that fight against cancer cells). Finally, gene modification can result in the secretion by cells of molecules not normally associated with those cells; for instance, lymphocytes could be made to secrete tumor necrosis factor, which could increase their anticancer activity. Although it was long thought that most applications of gene therapy would involve correction of inherited genetic metabolic abnormalities, today most applications attempt to improve cancer treatments and are being tested in patients for whom all standard therapies have failed.

Since disease can be seen as the absent or aberrant function of normal physiologic processes controlled by gene expression, opportunities to apply gene therapy principles are as varied as the diseases themselves. Even small gene changes can make a large difference, as is evidenced by the fact that only 1.6 percent of the human genome differs from that of the pygmy chimpanzees. Genes that can affect the function of human cells can come from other people or from virtually any other life form and thus the possibilities for genetic manipulation are extremely varied.

As with any new technology, gene therapy can be used responsibly, or it can be abused, and concern regarding the safety of gene transfer as well as the ethical and social issues related to its use has led to substantial controversy. Some thought that the use of viruses to introduce genes might result in unexpected hazards. Retroviruses most commonly used for human gene therapy come from viruses that can cause leukemia in mice, but have no known human hazard. Although these viruses have been engineered so that they cannot reproduce, it is possible that they could combine with similar viruses naturally present in people and in that way develop the ability to replicate. To avoid this troubling possibility, viruses are carefully engineered with sequences that inactivate any new virus that may emerge. Similarly, adenoviruses used in gene therapy have had the genes required for the creation of new adenovirus deleted.

WILL GENE THERAPY SOMEDAY ERADICATE ALL INBORN GENETIC DISORDERS?

Others have argued that tampering with the human genome crosses a boundary that should not be breached and that such changes could alter the course of human evolution. The natural course of human evolution could be altered, they say. Current therapies modify only somatic cells and do not effect germ cells in the ovary and testes that could result in gene transfer to offspring. But since the

Birth of an Era

The transfer of foreign genes into humans had a tumultuous debut. After years of laboratory research and experimentation in animals, investigators, Mike Blaese, French Anderson and Steven Rosenberg proposed a way to introduce bacterial genes into the immune lymphocytes of cancer patients in order to determine the traffic and survival of the lymphocytes in the body. This would be the first time that foreign genes are introduced into humans. On October 3, 1988, the presentation was made. Despite the fact that these patients had advanced cancer, there was great criticism about the possible dangers of using retroviruses to introduce foreign genes into cells. We argued that these cancer patients were in desperate need of improved treatment. Nonetheless, criticism mounted for deferred gene therapy, until finally, Bernard Davis, a highly respected Harvard scientist, made an impassioned plea. "It is virtually not possible," he argued, "to have more risks than certain death." In a press release from Dr. Wyngaarden's Office, Frank Young, the Director of the Food and Drug Administration, announced his approval of the protocol as well. It still wasn't over.

On January 30. 1989, at the next regularly scheduled meeting of the Recombinant DNA Advisory Committee (RAC), Jeremy Rifkin, a highly critical biotechnology activist, announced that he was filing a lawsuit in federal court to halt the experiment. Rifkin stated, "It is the first experiment in the world in which a foreign gene is to be placed into a human being. With this experiment, we begin the whole era of human genetic engineering... If we are not careful, we will find ourselves in a world where the disabled, minorities, and workers will be genetically engineered... We will be back next time and the next time!... We'll be back here every single time. You know we won't go away!" Handicapped people brought by Rifkin testified against the protocol as well. The lawsuit was based on a technicality that one of the RAC votes had been taken by mail instead of at a public meeting. The NIH quickly resolved this lawsuit and we received final permission to proceed. The first patient to receive a foreign gene was treated on May 22, 1989.

GENE THERAPY MAY SOMEDAY MAKE TRADITIONAL TREATMENTS, SUCH AS DRUGS, SURGERY AND RADIATION, OBSOLETE.

potential ability does exist to introduce genes into germ-line cells that could change future generations, there is now general agreement that germ-line gene therapy should not be attempted until we have a deeper understanding of how genes and developing cells interact. While there is now a great deal of support in the United States for gene therapy, the technique is illegal in many countries. There is, moreover, a general belief that gene therapy should not be used to modify traits that might enhance desirable, but not essential, characteristics—increased strength or greater height, for example.

Physicians have traditionally treated patients using external forces such as scalpels, drugs or radiation beams. The advent of gene therapy has introduced a new approach to treatment that provides the body with internal tools with which to replace defective natural functions or bolster defenses against disease. The future applications of gene therapy have the potential to profoundly change the practice of medicine as we know it today.

ARBITRARINESS OF RACE

Classification of humans into groups (races) on the basis of their physical appearance is an old exercise. Since the eighteenth century, many observers have proposed various schemes, with the number of races varying from three to many dozens. Charles Darwin was already aware of the central difficulty which made such efforts futile: a great number of intermediate groups form an almost perfect continuum. Rigorous methods based on a quantitative study of genetic (biologically inherited) characters were introduced by modern genetics, and they show that, without a doubt, all such racial classifications are arbitrary. There is a

Luca Cavalli-Sforza

very extensive genetic variation among individuals, and any human group, however small and isolated, shows its members differ one from the other to a major extent; which is remarkably high, even when compared with the differences observed at the global level, among major races. Thus, the racial purity is a myth. It was dear to social scientists of the nineteenth century, and has still been influential in the twentieth. It was accompanied by a belief in the superiority of the white race, as well as a conviction, despite a complete lack of evidence, that the decay of civilization was caused by the admixture of different races. The fact is that we do not know of any particular biological disadvantage affecting individuals whose parents belong to different "races," and in fact, to some extent there may be advantages in having a mixed parentage. Naturally, however, a combination of prejudice and cultural differences may put such individuals at a particular kind of disadvantage.

We know that biologically inherited differences between individuals are reflected in the DNA. How do they arise? They are part of the evolutionary process, and in particular, of the phenomenon called mutation. DNA passed to a child by a parent is a very faithful copy of that of the parent; it allows only a few errors, called mutations. DNA is made of a string of elements called nucleotides, which are of four types: A,C,G,T. The sequence of these nucleotides is responsible for all our inherited characteristics. In humans, it is made of three billion elements: each of us has two such sets of about three billion nucleotides, one received from the father, the other from the mother. If we compare them, we may find that about one nucleotide in every thousand is different in the two sets, paternal and maternal. If we compare one such set from an individual with that from a different individual from the same population, we commonly find about the same differences as between the paternal and maternal sets of an individual. If we compare sets from

111

individuals of two populations belonging to different "races," we would find only a light increase in the difference.

The effect of a mutation on the organism that carries it is often trivial and totally unnoticeable in physical appearance. But sometimes it may cause trouble, and much more rarely an improvement. The new type brought about by mutation can be passed to descendants. All the differences we notice between two individuals are the result of mutations which occurred in our ancestors, often a very long time ago, even a million years or more; or, at the other extreme, only in the last generation or a few generations ago. Mutation is a rare but unavoidable event. "Bad" mutations are eliminated by the process of natural selection: individuals who carry them have less chance of surviving and reproducing, and thus their DNA has a smaller chance of being represented in future generations. But "good" mutations may help develop fitness of organisms. Good and bad are usually

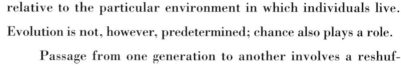

relative to the particular environment in which individuals live. Evolution is not, however, predetermined; chance also plays a role.

Passage from one generation to another involves a reshuffling and a random sorting of the individual types of DNA, as in a card game; a process called random genetic drift. It may be especially effective when a small group of individuals migrates from its place of birth and cuts off or reduces its ties with its origin. The world has been settled by many such migration episodes, and this movement was in part responsible for the generation of racial

differences. But the migrants also had to adapt to the different environments, from the equator to the arctic, where they settled. By this mechanism, too, migrations have caused differentiation. But there is always a little bit of "creeping" migration, every time and everywhere, much of it because husband and wife are born in different places. Usually parents live very near each other, even today when transportation has become so much easier. This makes neighboring populations very similar to one another, and even more similar the closer they are geographically, a proximity that generates a trend toward continuity of genetic variation which hampers racial classifications. In this way, migration can have a homogenizing effect.

Four evolutionary factors—mutation, selection, drift, migration—explain much of the evolution we observe. We have used data on the genetic diversity thus generated to reconstruct the history of humankind. Although the details have yet to be confirmed, there is substantial agreement between the results of our historical analysis based on genetic data and facts gleaned from a variety of disciplines, such as paleoanthropology, archaeology, and linguistics. Each provides some evidence, some pieces in the mosaic.

What we can guess at, based on the general picture, suggests that ancient humans similar to modern people had a demographic explosion generated by some advantage, like the ability to make better tools, or to communicate better with a more advanced language. Their origin was probably in Africa. Humans then migrated to Asia and from there, probably in the last 50-60,000 years, to Australia, Europe and the Americas. Whether and to what extent they mixed with earlier inhabitants is not yet known. Although some object, the impression is that the contribution of earlier inhabitants—Neanderthals, for example, who lived in Europe until some 30-35,000 years ago—is marginal or absent.

In the last ten thousand years, climate, flora and fauna changes and a condition of relative population saturation with respect to the means then available of procuring food (hunting, gathering and fishing), stimulated the transition from food collection to food production. This trend generated a technological revolution which in turn propelled local population explosions. Such events occurred in many parts of the world, but most commonly in temperate climates. These people often expanded to neighboring favorable regions. The geography of human genes shows traces of these expansions. As the people took their culture—technology, language, customs—with them, we see unexpected parallels between different aspects of human  diversity. For example, one finds coevolution of languages and genes in unexpected places. The transmission of culture follows laws other than those of genes, and such coevolution may seem strange to us, as we are used today to extremely fast communication, and therefore a potentially rapid diffusion of culture. In ancient times, cultural transmission must have been mostly passed from parent to child, and largely limited to social group. Cultural transmission was similar to genetic transmission, making gene-culture coevolution unavoidable. The phenomenon explains, among other things, strong similarities between genes and language.

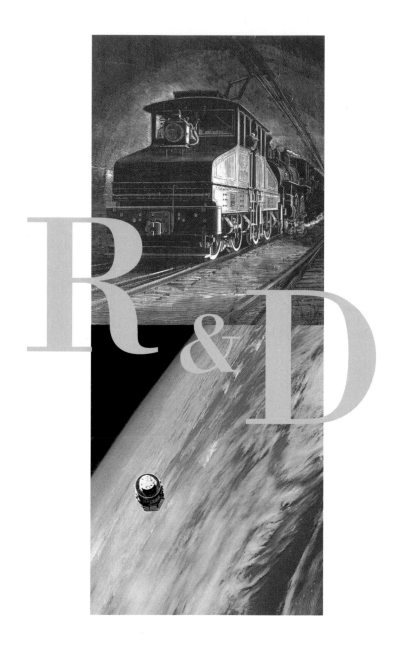

GOVERNMENT-FINANCED RESEARCH & DEVELOPMENT

While the standard mythological image is that of America's industrial prowess being nurtured by its great private entrepreneurs (Ford, Edison, Rockefeller, Carnegie) alone, the reality is much more complex. Government-financed industrial policies have always played an important role in United States' economic success.

America's first, and perhaps still greatest, contribution to industrialization, was the invention of Eli Whitney's interchangeable parts, a project funded by the U.S. War Department to find a way to make better and cheaper muskets.

Development moved west, first with the Erie Canal and then with the transcontinental railroads—both requiring substantial government backing. The land-grant college system, built to enhance agriculture with technology and

Lester C. Thurow

electricity, was brought to American agriculture with the Rural Electrification Administration. The result was an agricultural exporting powerhouse that had not previously existed.

The United States became the world's leader in science and engineering under the impetus of the spending for military R&D during the Second World War. Before then, one went to Germany to find the world's leaders in science and technology. Post-World War II leadership in civilian aircraft manufacturing grew out of the technologies and economies of scale learned in military aircraft production. Government research and purchases fueled the startup of the civilian computer industry. Teflon moved from the space program to the kitchen. Biotechnology exists, and is almost an American monopoly, because the U.S. government was willing to pour billions into research decades before there was even a hint of its commercial potential.

The world, however, has been evolving rapidly in recent years, and today, America is in two senses at a turning point in its history of industrial policies. What has been episodic will have to become systematic. First, with the end of the Cold War, the scope and size of the military R&D system is being reduced. To match R&D spending of our leading economic competitors (Germany and Japan), a new civilian system of R&D will have to be built to replace it. Second, the world is leaving an era where comparative advantage depends upon raw materials or the availability of capital, and is entering an era of brain-power industries where comparative advantage is human.

All social systems have strengths and weaknesses. Capitalism is no exception. It weakness is its myopia—its short-sightedness. Using any reasonable interest rate,

◄

the discounted net present value of a dollar in earnings eight or nine years in the future is near zero today. As a result, capitalist firms won't invest in projects with long payout periods. Yet major infrastructure projects and fundamental R&D usually don't pay off in eight or nine years.

As a result, governments must play a central role in financing industrial research and development and in financing the infrastructure projects that have to be built ahead of the markets that they will ultimately serve. Private firms will only build the information highway when it is clear that there are sellable services that can use it. The U.S. Defense Department fueled the Internet, now the model for the information highway.

Building a civilian R&D system to replace the current military system demands more fundamental thinking about industrial policies. To know where to spend one's research money, one has to have a sense of which are the hot civilian technologies that might pay off in big breakthroughs and new high-wage-employment-generating products. Just so, the military had to have a sense of which military items it needed and which technologies would be needed to meet those objectives. If anything has been learned from R&D management, it is that you cannot just give researchers money and tell them to do "good things." There has to be a vision of what one wants or needs. By the same token, there has to be accountability—that is, a sense of when failure has occurred.

Flat-Screen Race

▲

GOVERNMENT R&D
HELPS U.S. FIRMS WITH
TECHNOLOGY THAT IS
USED IN THE DEVELOP-
MENT OF IMPROVED
FLAT-PANEL DISPLAYS.

In April 1994, the Pentagon announced a $1 billion, ten-year government R&D investment program to help American companies catch up on flat-panel displays, a technology now dominated by the Japanese. Participating private firms had to match government R&D funds and commit to building four state-of-the-art manufacturing plants. The goal is to push both product and process development. The military did not want to be completely dependent on foreign suppliers, but a prime objective is creating a competitive American presence in large civilian markets for flat-panel displays.

Flat-panel displays are now used in aircraft cockpits, laptop computer displays, and goggles for virtual-reality video games. If they can be scaled up in size economically, they will likely be the future screen of choice in almost every application where standard TV displays are now used.

The advantage of being a follower is that one already knows that there is a large commercial market for the product. The Japanese simply own it. The question is, can it be taken away from them?

Such a vision intrinsically requires a willingness to invest in the long run. The Star Wars military R&D program began even though its proponents did not believe that it could be realized within a quarter of a century. While we clearly have been willing to make long-run investments in military and clinical areas, it is not clear that we are willing to make the same long-run investments in conventional civilian technologies. That part of the economy simply rode the spin-offs from the military R&D system. Those fortunate spin-offs will have to be replaced by a systematic search for equally valuable processes and products.

Consider any list of what are expected to be the most rapidly growing high-wage high-rate-of-return industries of the next few decades. Industries such as microelectronics, biotechnology, the new materials industries, telecommunications, civilian aircraft production, machine tools plus robotics, and computers plus software are all brain-power industries. All are geography-free. They could be anywhere. They will be in those places where someone organizes the brain power.

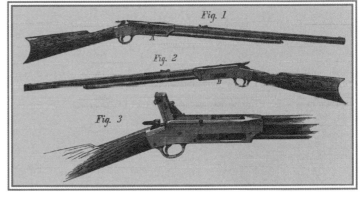

In contrast, natural-resources industries of the past had fixed geographic homes. Before the development of world capital markets, capital-intensive products were built in rich nations and labor-intensive products were built in poor countries. Today, when everyone effectively borrows in New York, London and Tokyo, there are no countries which are intrinsically capital- or labor-intensive.

The industries of the future will not emerge merely by picking winners and losers. The U.S. military-industrial-R&D complex was not in the business of picking winners and losers. What it did was to articulate a mission, and then initiate a complex set of interactions between military demands, an industrial base and an R&D complex; those interactions spanned government, industrial and university laboratories. In a similar fashion, a successful civilian industrial policy depends upon an exchange of information and a web of investments, partly public and partly private. In a world of human-made comparative advantage, others will be seeking to create the conditions that allow them to gain their share, or more than their share, of the high-wage, high-rate-of-return growth industries of the future. The great example of challenge is, of course, the European government-owned consortium, Airbus Industries, that seeks to challenge American dominance in civilian aircraft manufacturing. What is the correct response?

▲

THE U.S. WAR
DEPARTMENT FUNDED
PROJECTS TO MAKE
BETTER AND CHEAPER
MUSKETS.

AGE OF UNCERTAINTY

USING DATA FROM A NATIONAL NETWORK of weather stations they set up during World War I, Norwegian geophysicist Vilhelm Bjerknes and his son, Jacob, found modern meteorology when they show that the atmosphere is made of large, well-defined air masses that display vastly different temperatures. The father-son team adopt the wartime term "front" to demarcate the boundaries separating one air mass from another.

21

CANADIAN PHYSICIANS Frederick Banting and Charles Best tie off several dogs' pancreatic ducts and, seven weeks later, extract a critical solution from each degenerating organ. They have isolated the hormone insulin, which will soon provide a successful treatment for the debilitating disease diabetes.

22

THE SO-CALLED PRIMORDIAL SOUP theory of Russian biochemist Alexander Oparin is the first sound scientific hypothesis rivaling biblical creationism to explain the origin of life. Oparin suggests that life developed through the slow accumulation of organic compounds in an oxygen-free atmosphere sparked by natural sources of energy, such as lightning.

AN ARCHAEOLOGICAL EXPEDITION led by George Herbert, fifth Earl of Carnarvon, and British archaeologist Howard Carter unearths the riches of ancient Egypt when they find the tomb of 21-year-old pharaoh Tutankhamen, who died in 1352 B.C. The tale of the "Pharaoh's Curse" begins when Lord Carnarvon dies only five months later.

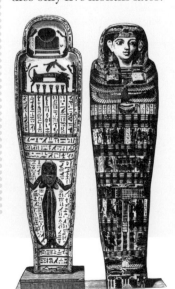

23

FRENCH PHYSICIST LOUIS DE BROGLIE suggests that if electromagnetic energy can behave more like particles than waves, matter can behave more like waves than particles. He hypothesizes that particles such as electrons have an associated wavelength determined by their momentum.

25

AUSTRALIAN-BORN SOUTH AFRICAN anthropologist and surgeon Raymond Dart announces the discovery of a small, previously unknown hominid skull in a South African limestone quarry. It is the first recognized Australopithecus, an upright walking creature more similar to humans than to apes.

THE SO-CALLED SCOPES MONKEY trial is held in Dayton, Tenn. High-school science instructor John Scopes is convicted of violating a state law prohibiting schools from teaching Darwinian evolution.

26

THE FIRST LIQUID-FUEL ROCKET, only four feet long and six inches in diameter, is shot 200 feet into the sky above Worcester, Mass. The rocket's inventor, U.S. physicist Robert Goddard, believes it is the first step to a human voyage to the moon. The press responds with ridicule.

27

GERMAN PHYSICIST WERNER HEISENBERG advances his famous uncertainty principle—declaring that it is not possible to know simultaneously both the momentum and position of a subatomic particle.

THROUGH CLOUDS, SLEET and darkness, U.S. aviator Charles Lindbergh is the first solo pilot to fly nonstop across the Atlantic Ocean. Lindbergh's voyage to Paris in his celebrated monoplane *The Spirit of St. Louis* takes 33 hours, 29 minutes.

28

WERNER HEISENBERG, Austrian physicist Erwin Schrödinger, and British physicist Paul Dirac reformulate Newtonian mechanics into quantum mechanics. Particles have a quantum state, a combination of position and velocity. Quantum mechanics and Einsteinian relativity are the two great theoretical foundations of modern physics.

PAUL DIRAC PROPOSES the existence of antiparticles—particles identical to protons and electrons but with opposite charges. When a particle collides with its antiparticle counterpart, he hypothesizes, they annihilate each other, leaving only energy. Four years later, discovery of the positron—an antiparticle

THE SPIRIT OF SAINT LOUIS
▼

electron—is announced.

A PENICILLIUM NOTATUM MOLD mistakenly contaminates a staphylococcus bacterial culture in the London laboratory of British bacteriologist Alexander Fleming. To his surprise, the bacteria surrounding the mold mysteriously die. Fleming names the mold's powerful toxin penicillin. As an antibiotic, it will save the lives of thousands of soldiers in World War II and become available in civilian hospitals by 1944.

29

PHOEBUS LEVENE, who discovered ribose nucleic acid 20 years earlier, finds another sugar, deoxyribose, in nucleic acids. Although he does not yet know the significance of his finding, he has discovered deoxyribonucleic acid or DNA.

BILLIONS OF DOLLARS in paper profits vanish when the U.S. stock market crashes on October 24. The worldwide Great Depression has begun.

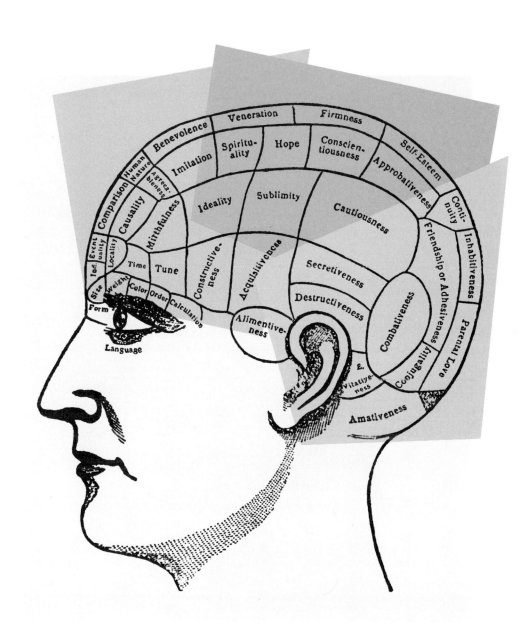

SOLVING FOR g
AND BEYOND

For centuries, observers have pondered the nature of intelligence. Where does intelligence come from? Is it an unchangeable aspect of human nature? Or can it be enhanced? By what means can individuals be judged intelligent? Although such questions are ancient, only in the last 150 years or so have they been empirically studied.

Until the 1800s, many maintained that the human mind was divine in origin and separate from the human body. But by the middle of the nineteenth century, physiologists and anatomists had demonstrated that thought was indeed linked to

**Howard Gardner
and
Mindy L. Kornhaber**

mortal flesh. By calculating the time it took for individuals to press a button after receiving physical stimuli on different parts of their leg, German physicist and physiologist Hermann von Helmholtz determined

that the nerve impulses travelled at measurable, rather than divine or spontaneous speeds. A bodily basis for intelligence also came from demonstrations by Paul Broca and others that the ability to speak was impaired when particular regions in the brain's left hemisphere were damaged.

The scientific study of intelligence was also spurred by Charles Darwin's theory of evolution. Darwin argued that human intelligence lies on a continuum with—rather than being distinct from—animal intelligence. According to Darwin's theory, useful variations that occur within species tend to be inherited. This notion inspired Darwin's half-cousin, Francis Galton, to found and advocate eugenics, or the selective breeding of people to enhance the human population.

◄

SCIENTISTS HAVE
LONG STUDIED
THE LINK BETWEEN
MIND AND BODY.

To help determine which people should be encouraged or discouraged from reproducing, Galton invented the mental test. Galton's tests, however, were not like those commonly used today. Since he believed that intelligence stemmed from powers of sensory discrimination, his tests asked individuals to detect small differences in such things as smell, visual stimuli, or weight.

By the turn of the twentieth century, sensory discrimination tests like Galton's had been rejected in favor of tests developed in France by psychologists Alfred Binet and Theodore Simon. These tests sought to measure so-called higher order skills, such as memory and comprehension, and they consisted of sets of questions geared to children of different ages. A child's test results yielded one number, the mental age. This number referred to the most advanced set of questions that a child could adequately address. For instance, if an 8-year-old could answer most of the questions intended for a 10-year-old, her mental age was 10. From this number, the first IQ scores were soon derived by dividing a child's mental age by her chronological age and multiplying by 100. Thus, the child discussed above would have an IQ of 125.

About the time Binet and Simon were conducting their research, British psychologist Charles Spearman was also investigating children's test results. Spearman found that children's scores on quite different kinds of tests, ranging from school subjects to pitch discrimination tasks, were all positively correlated. That is, a high score on one measure was associated with a high score on the others; a low score on one measure was associated with low scores on the others. From such findings, Spearman argued that the basis for intelligence was a single, underlying ability he called general intelligence, or g.

Throughout this century, g has played a very prominent role in research on intelligence. Indeed, scores from tests of the sort developed by Binet and Simon soon came to be viewed as reliable measures of g.

Some research involving g has been aimed at understanding the relationship of g to other abilities. Is g solely responsible for intelligence or are some other factors involved? To answer this question, statistical tools are applied to the results from batteries of intelligence tests. These analyses typically show that scores on different kinds of mental tests were positively correlated, supporting g. However, they also reveal that test results cluster together around certain kinds of abilities—indicating that in addition to g there may be factors for spatial, mathematical, linguistic, and other skills.

Research since the 1960s has involved efforts to find the physiological origins of g. Some of these investigations have uncovered positive correlations between IQ scores and reaction times. This has led researchers such as British psychologist Hans Eysenck and U.S. psychologist Arthur Jensen to the controversial finding that intelligence is related to the speed or efficiency by which information is transmitted within the nervous system.

Current research in intelligence is proceeding in a number of directions. One enduring theme is the extent to which differences in intelligence within a population are inherited, and whether the genetic underpinnings of intelligence can be identified. A second theme is the optimal means for conceptualizing intelligence: as a single ability, a set of hierarchical abilities (including spatial and other skills) or as heterarchical (comprising a number of independent intellectual factors). A third theme, pioneered by psychologists Robert J. Sternberg and

Earl Hunt, entails analyzing the component processes involved in accomplishing intellectual feats such as solving analogies.

Some psychologists call for a more radical reexamination of the conceptualization and measurement of intelligence. Their research may rely on biological and cultural evidence that allows them to uncover mental abilities or "intelligences" not ordinarily measured or recognized by conventional methods. Other researchers assert that laboratory and test settings, where intelligence is traditionally investigated, lack the resources that commonly support our everyday intellectual activity. These resources include people, books, magazines, and computers and other tools that help us to organize and manipulate information.

To many of these investigators, intelligence is not defined solely by what's inside our heads. Rather, it is said to be distributed—existing in complex interactions among our mental abilities and available resources. For these researchers, simply measuring intelligence is less important than seeking to devise better systems of work and education that will allow more of us to tackle intelligently the problems we encounter in the course of our lives.

Child's Play

How do newborns ultimately attain adult mental abilities? This and related questions are pursued by developmental psychologists, among whom the undisputed giant was the Swiss researcher Jean Piaget.

Piaget's investigations of his own children yielded many important findings. One was that infants develop the ability to represent objects mentally the way older children do. Young infants continue to look for an object in its original location even after they have seen someone "hide" it elsewhere.

Piaget proposed two linked mental processes to explain how infants' minds gradually develop into adult minds.

Assimilation, he argued, enables infants to fit information from the surrounding environment into an existing mental framework or scheme. Accommodation, in turn, entails altering mental schemes to account for new information from the environment.

While assimilation and accommodation help explain the development of intelligence, recent research indicates that newborns are more capable even than Piaget believed. It appears that not only interactions with the surrounding environment, but inborn mental structures, account for the remarkable mental development young children display.

WE KNOW MORE THAN WE LEARN

A striking property of any natural language is that children acquire infinitely more of it than they experience. So there is more to language acquisition than simply mimicking what one hears in childhood.

Consider certain subtleties that people are not usually conscious of. The verb *is*, for example, may be used at times in either its full form or its reduced form. We may say *Kim is happy* or *Kim's happy*. However, most of us are unaware that for certain usages, *is* will never be reduced. Note, for example, the underlined verbs in *Kim is happier than Tim is* or *I wonder what that is up there*. One just knows not to use the reduced form here, but how did we come to this knowledge? As children, we are not usually instructed to avoid the reduced form in certain places. Still, children typically acquire the ability to use the verb forms in the adult fashion, regardless of intelligence or educational level. They attain this ability early in their linguistic development, and they do not "experiment" with the non-occurring forms as if testing some hypothesis, in the way that they do experiment with verbs like *goed* and *taked*.

David Lightfoot

In another example: consider that pronouns like *she*, *her*, *he*, *him*, and *his* generally may refer back to a noun previously mentioned in a sentence. Now read sentences 1a through 1d. Notice that 1d can only be understood as referring to two men; Jay and somebody else. Here, and only here, the pronoun may not refer to Jay.

1A Jay hurt his nose.

1B Jay's brother hurt him.

1C Jay said he hurt Ray.

1D Jay hurt him.

So, a generalization emerges, and is accepted, to the effect that a pronoun may refer to a preceding noun except under some very precise conditions such as those that exist in sentence 1d. But again, how did we all learn the right generalization, and particularly the right "except" clause?

Recall the nature of our childhood experience: we were exposed to a haphazard set of linguistic expressions. We heard various sentences containing pronouns;

EVEN IN THE EARLY TWENTIETH CENTURY, SCIENTISTS WERE CAPABLE OF INVENTING DEVICES (*LEFT AND FAR LEFT*) THAT COULD *MIMIC* SPOKEN LANGUAGE.

sometimes the pronoun referred to another noun in the same sentence, sometimes to a person not mentioned there. Problem: because we were not informed about what cannot occur, our childhood experience provided no evidence for the "except" clauses. That is, we had evidence for generalizations like "is may be pronounced z" and "pronouns may refer to a preceding noun," but no evidence for where these generalizations break down; our linguistic experience was not rich enough to determine the limits to the generalizations. This is the problem of the poverty of the stimulus, and the examples above are two small illustrations of the form that it takes in language.

There are two easy solutions to the poverty-of-stimulus problem, but neither is adequate. One is to say that children do not overgeneralize; that is, they do not produce the reduced *is* in the wrong place, or use the pronoun in 1.d incorrectly to refer to Jay, because they never hear such forms. In other words, children acquire their native language simply by imitating the speech of their elders. We know this solution is not correct because everybody constantly says things that they have never heard. A variant on that solution might be that children learn not to say the deviant forms because they are corrected by their elders. But that view also cannot be right, for several reasons. First, it would take an acute observer to detect and correct each error. Second, observation shows that where correction is offered, children are highly resistant. Third, in the examples discussed, children do not in fact overgeneralize and therefore parents have nothing to correct. So the first "easy" solution to the poverty-of-stimulus problem is to deny that it exists, to hold that the environment is indeed rich enough to provide sufficient evidence for where the generalizations break down.

The second "easy" answer denies that there is anything to be learned, and holds that a person's language is fully determined by genetic programming. This answer is refuted by the fact that people speak differently, and many of the differences are environmentally induced. There is nothing about my genetic inheritance that makes me a speaker of English; if I had been raised in a Dutch home, I would have become a Dutch speaker.

The two "easy" answers attribute all language acquisition either to the environment or to genetic inheritance. Neither position is tenable. Rather, language emerges through an interaction between our genetic inheritance and the linguistic

▶ THIS SYNTHETIC APPARATUS FOR FORMING VOWEL SOUNDS WAS FEATURED IN *SCIENTIFIC AMERICAN* ON OCTOBER 5, 1901.

environment to which we happen to be exposed. English-speaking children learn from their environment that the verb "is" is pronounced "iz" or "z," and genetic principles prevent the reduced form from occurring in the wrong places. Likewise, children learn from their environment that he, his, etc. are pronouns, while genetic principles tell where pronouns may not refer to a preceding noun. The interaction of environmental information and genetic principles accounts for how the relevant properties emerge in an English-speaking child.

Since children are capable of acquiring any language to which they happen to be exposed between infancy and puberty, the same set of genetic principles which account for the emergence of English must also account for the emergence of Dutch, Viet-

our father who art in heaven, hallowed be thy name. thy kingdom come. thy will be done in earth as it is in heaven. Give us this day our daily bread. and f-

namese, Hopi or any other of the thousands of languages spoken by human beings. This plasticity imposes a strong empirical demand on hypotheses about the linguistic genotype; the principles postulated must be open enough to account for the vast amount of variation among the world's languages. The fact that people develop different linguistic capacities, depending on whether they are brought up in Basel, Bali, or Baltimore, provides a delicate tool with which to refine claims about the nature of the genetic component.

So, there is a biological entity, a mental organ, which develops in children along one of a number of paths encoded in the genes. The language organ that emerges is represented in the brain and plays a central role in the person's use of language, whether it is used for speaking, listening, writing poems or solving crossword puzzles. We have gained some insight into the nature of people's language organs by considering a wide range of phenomena: the developmental stages that young children go through, the way language breaks down in the event of brain damage, the manner in which people analyze incoming speech signals and more. At the center is the biological notion of a language organ.

If asked to describe quite generally what is now known about the linguistic genotype, I would say that it is plastic (consistent with speaking Japanese or Quechua), modular and uniquely computational. By "modular," I mean that it consists of various modules which interact in specified ways, and which indeed are specific to language. Research has disproved the notion that the mind possesses only general principles of "intelligence" under which are mixed together all kinds of mental activity. One module of innate linguistic capacity contains abstract structures which are compositional (consisting of units made up of smaller units) and which fit a narrow range of possibilities. Another module holds the ability to relate one position to another within these structures, and those relationships are narrowly defined. For a taste of the computational mechanisms involved, consider another reduction process, whereby *want to* is often pronounced wanna: *I want to be happy* or *I wanna be happy*. Here, the to may be attached to the preceding *want* and the phrase may be reduced to give the *wanna* form. But an intervening understood element (x) blocks the reduction process. For example, *who do you want to see?*, which has the logical structure of 2a (corresponding to the statement you want to see Jill), may be pronounced with the *wanna* form.

A TURN-OF-THE-CENTURY EXAMPLE OF MANOMETRIC FLAMES OF FRENCH VOWEL FORMATIONS WAS AN EARLY FORM OF SPEECH RECOGNITION.

2A whox do you want [to see x].

However, when an understood element intervenes between *want* and *to*, as in expressions like *who do you want to go?*, which has the logical structure of 2.b, *wanna* does not usually occur.

2B whox do you want [x to go].

Similarly, the reduced *is* is attached to the *following* word, contrary to what the spelling convention suggests. If there is no following word, the reduced form does not occur: *Kim is happier than Tim is*. And an intervening understood element also blocks reduction: *I wonder what that is x up there*. Now we have an answer to the problem sketched in the second paragraph of this essay. There are further restrictions on reduced forms, but the restrictions are general and are formulated in such a way that no attachment process of the appropriate class in any language may operate across an understood element of the appropriate class. Defining exactly what is meant by "the appropriate class" is where much of the empirical work of linguistics lies, and linguists spend much of their time analyzing a wide range of languages.

For centuries, people have held that organisms grow according to the instructions of an internal regulatory program. As a result of twentieth-century work in molecular biology, we now understand much about how that regulatory mechanism works, how it is transmitted, how it can be amended, and so on. But the idea that there must exist an internal regulatory program long antedates recent work. For hundreds of years scientists theorized that human sperm contained a perfect, complete miniature person, a "Russian doll" or "homunculus," which simply grew bigger over time. This was Preformationism, which was a widely accepted theory in the eighteenth century. Earlier, Plato held that the basic ideas and elements of thought were innate and that at birth we drank of the River Lethe, the river of forgetfulness, which relegated this information to the subconscious. We no longer hold such ideas, but they demonstrate that the compelling question of how living things develop has always been the basis for much theorizing, and that people have long thought that the development is in some way internally directed. The reason for this belief has always been that environmental influences are too poor to determine certain features of the make-up of a mature organism.

Recently, theoretical developments have brought about an explosive increase in our knowledge of human languages. Linguists can now formulate interesting hypotheses, and can account for broad ranges of phenomena in many languages with elegant abstract principles. In some ways, work on human grammars has paralleled work on the visual system and has reached similar conclusions, particularly with regard to the existence of highly specific computational mechanisms. In fact, language and vision are the areas of

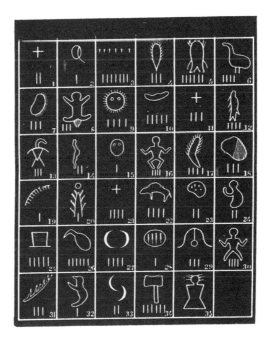

cognition that we know most about. Much remains to be done, but we can show how children acquire certain elements of their language organs through exposure to what is essentially a disorganized and haphazard set of simple utterances; for these elements we have a theory which meets basic requirements. Eventually, the development of language in a child will be viewed as similar to the growth of hair: just as hair emerges with a certain level of light, air, and protein, so, too, a biologically regulated language organ necessarily emerges under exposure to a random speech community.

VERY INTENSE LIGHT

Lasers produce light with remarkable qualities—such light is usually of a very pure frequency or wavelength and is emitted in a highly directional beam, which also allows it to be focused to a point as small as one fifty-thousandth of an inch. Lasers may also give very intense power, or pulses as short as one hundred-thousandth of a billionth of a second. Laser technology combines the fields of optics with the precise control and flexibility of electronics, and has a wide range of applications in both science and technology.

It was James Clerk Maxwell in the nineteenth century who showed that light is an electromagnetic wave and similar to many other now-familiar types of radiation

Charles H. Townes

from the long waves of radio to the successively shorter ones described as microwaves, infrared, visible light, ultraviolet, and x-rays. Albert Einstein first recognized that light also occurs in discrete bundles of energy, now called photons, and that if light can be absorbed by materials, the atoms or molecules in the materials must not only be able to produce light when heated, but can also produce photons of light by stimulated emission.

If an atom is in its lowest energy state, a photon of light may be absorbed by it, elevating it to a higher state of energy. It may then spontaneously drop back to the lower energy state, emitting a photon of light by "spontaneous radiation." Or, if light of just the right wavelength or photon energy passes near the atom while it still has this energy, it may stimulate the atom to release the energy to the light wave, thus building up the energy of the light wave or amplifying it by "stimulated emission."

This amplification is not usually seen because, under normal conditions, a collection of atoms always has more members in a lower state of energy than in a higher one, and since each of the lower-energy atoms absorbs light with the same probability that the higher-energy ones give up energy to the wave, the net effect is that the wave loses energy or is absorbed. However, if a collection of atoms or molecules is given energy in special ways to produce the unusual condition where there are more in a higher state of energy than in a lower one, then the sum of processes of absorption from the lower state and stimulated emission from the upper one will result in amplification, or intensification of a light wave. This is laser action, or *l*ight *a*mplification by *s*timulated *e*mission of *r*adiation, which produces light with characteristics never before seen by humans.

◀

MORE THAN ONE MILLION MICROLASERS CAN BE FOUND ON THIS SEMICONDUCTOR CHIP ABOUT SEVEN MILLIMETERS WIDE BY EIGHT MILLIMETERS LONG; A SMALL ARRAY OF MICROLASERS IS SHOWN.

The field of study that includes lasers is sometimes called quantum electronics, because it deals with electromagnetic waves, as does electronics, but in ways where the "quanta," or photons, are important. Furthermore, the science emerged from studies of molecular absorption of microwaves, a field normally recognized as electronics. Fundamental physical research on absorption of microwaves by molecules and the eagerness of physicists to obtain still shorter waves to extend such studies led C.H. Townes at Columbia University in 1951 to the idea of producing short waves by stimulated emission. He, together with J.P. Gordon and H.J. Zeiger, completed a successful system in 1954. A beam of ammonia molecules was sent into a cavity, with higher-energy molecules selectively deflected into the cavity by an electric field. Inside the cavity, microwaves bounced back and forth, stimulating the molecules to give up their energy and becoming amplified. This effect produced what was named a *maser*, for *m*icrowave *a*mplification by *s*timulated *e*mission of *r*adiation. It was an exciting device, and was adapted by many physicists and engineers to other forms, including masers made of solid materials. Masers produced very accurate "atomic clocks" because of their pure frequency, and were also used to make amplifiers which were more sensitive by at least a factor of ten than any other microwave amplifiers previously available. The maser is now used both for frequency standards and as an amplifier for space communications.

In 1958, Townes and Arthur Schawlow, then at the Bell Telephone Labs, showed how the maser idea could be extended to very much shorter wavelengths, even to waves as short as light. Two mirrors faced each other so that light could bounce back and forth between them, passing through atoms which were excited by illumination from along the sides of this path. With the correct mirror dimensions, only one wavelength of light would be amplified, and that so strongly that it would come through one of the mirrors, which was partially transparent, as an intense and highly directed beam.

Many laboratories set about trying to construct an "optical maser," now called a laser. The first to succeed was Theodore Maiman in early 1960 at the Hughes Laboratory. He used a rod of ruby with reflectors on each end and excited by a photoflash tube. This device produced intense pulsed beams of red light. Soon after, many types of lasers were introduced. Some were produced by gases excited by light; others by gases excited in an electrical discharge; still others by solid semiconductors excited by electrical currents. Some resulted from specially arranged chemical reactions, and one popular type uses liquid dye molecules excited by another laser.

The intensity of laser light is extremely useful in a variety of industrial and scientific purposes. It can cut and weld materials precisely and efficiently. Lasers can easily evaporate the hardest known materials, such as diamonds, or by very quick heating can provide remarkable hardening of bearing surfaces while allowing the interior of the material to remain cool and malleable. The light is intense enough to produce harmonics, of double the frequency. Lasers of modest and carefully controlled power can be powerful medical and surgical tools, particularly effective on human eyes or in other areas where light can penetrate.

One of the most direct uses of the laser's purity of frequency and high directivity is in measuring distances. The 240,000-mile distance to the moon has been measured this way to a precision of about one inch. The standard of length is now defined by the wavelength produced by a particular laser. Measurements of time have also been improved by quantum electronics. Time is now defined by an atomic clock using a beam of cesium atoms, and perhaps the most precise clock for moderate lengths of time is the hydrogen maser, producing oscillations at a frequency characteristic of hydrogen atoms.

Of the wide variety of industrial or technical applications of lasers, the transfer and storage of information by laser light is perhaps the most pervasive. Lasers are used in check-out devices in grocery stores, for high-fidelity recording or readout of music and information in compact disks, for laser printers and for efficient long-distance communication. Since the frequencies produced by lasers are very high, several hundred thousand billions of cycles per second or higher, they have enough bandwidth for a very high rate of transmission of information. In principle, the information from all TV, telephone, and radio transmissions in the world could be simultaneously transmitted on one light beam. At present, bandwidths as high as about 10 billion cycles per second are practical, although much narrower bandwidths are generally used. But even the presently relatively modest exploitation of the laser's potential has produced a revolution in communication. Modulated laser light transmitted on optical fibers is used particularly for long-distance multiple-channel transmission of information, including transoceanic communication.

Although lasers are now in widespread use, quantum electronics is still young and its applications continue to be explored. Directions of development include work towards useful x-ray lasers and powerful lasers to produce nuclear fusion. But probably much more important are the many everyday uses of lasers which are steadily increasing. Many of these are based on cheap and rugged solid-state lasers, which are now the most common variety.

MULTIPLICATION AND DIVISION

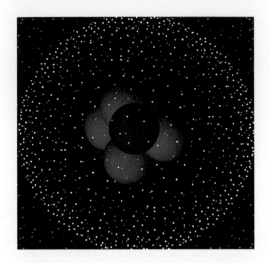

◀

COMPUTER-
GENERATED MODEL
OF BERYLLIUM
ATOM, SHOWING
NUCLEUS AND
ELECTRON ORBITALS.

1930

**AT THE ANNUAL
MEETING** of the American
Chemical Society, DuPont
chemist Thomas Midgley,
Jr., introduces an aston-
ishing new chemical
refrigerant. To demonstrate
its safety and nonflamma-
bility, he inhales the
gas and then blows out a
candle. The miraculous
chemical is CFC-12, the
first of the chlorofluorocarbons (CFCs), and
inspires the famous DuPont
slogan, "Better living
through chemistry."

31

**TO STUDY THE
STRATOSPHERE** first-
hand, Swiss physicist
Auguste Piccard constructs
a sealed gondola tethered
to a balloon and uses it
to ascend to a height of
nearly 10 miles—at that
time the greatest altitude
reached by a human.
(In 1960 his son Jacques
descends in a deep-sea
diving machine designed
by Auguste Piccard to
35,800 feet beneath the
Pacific Ocean. As of publication of this book, no
human has yet gone deeper).

32

**ENGLISH PHYSICIST
JAMES CHADWICK**
knocks uncharged particles
from a beryllium nucleus
and discovers neutrons,
which soon become impor-
tant for initiating nuclear
reactions. Werner
Heisenberg immediately
suggests that the atomic
nucleus, whose composition
has eluded physicists
since the discovery of the
electron, is made of protons
and neutrons.

**GERMAN ELECTRICAL
ENGINEER** Ernst Ruska
develops the first electron
microscope.

35 **GERMAN ZOOLOGIST KONRAD LORENZ** describes imprinting, the process by which a newborn bird takes the first large object it sees to represent its species. In experiments he finds that while such birds ordinarily imprint on their mother, another bird, a human or even a balloon can successfully be substituted. The science of ethology, or animal behavior, has begun.

RADIO DETECTION AND RANGING—radar, for short—is developed by Scottish physicist Robert A. Watson-Watt. He invents devices that will emit and detect reflected microwaves.

33 **VITAMIN C IS THE FIRST** of many vitamins successfully synthesized in the laboratory. Diseases caused by nutrient-deficient diets soon decline throughout the industrialized world.

AUSTRIAN-BORN NAZI LEADER Adolf Hitler becomes chancellor of Germany. He quickly rebuilds the military, eliminates opposition parties and begins persecution of the Jews. Both Germany and Japan withdraw from the League of Nations.

38 **A BONY PREHISTORIC FISH** called the coelacanth, believed to have been extinct since the age of dinosaurs, is caught by a trawler off the coast of South Africa. The coelacanth has fins attached to fleshy lobes rather than directly to its body and is likely a close evolutionary relative of the first amphibians.

39 **SWISS CHEMIST PAUL MÜLLER** discovers dichlorodiphenyltrichloroethane, a compound that appears to poison insects but not other forms of life. DDT is the first of many pesticides that initially succeed in controlling disease-carrying, crop-damaging insects but, through overuse, foster resistant insect strains and leave environmentally hazardous byproducts.

GERMAN PHYSICIST OTTO HAHN causes a fission reaction when he bombards uranium with neutrons. Other scientists confirm that fission can cause chain reactions that are self-sustaining. Many now realize that an atomic bomb is possible, and Albert Einstein writes a letter to President Roosevelt urging him to develop a nuclear bomb before Hitler can.

0101010100 1100

01 10

00 01

01 10

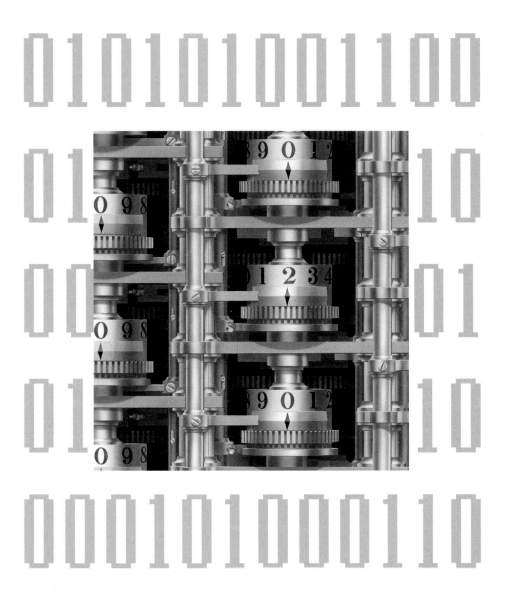

0001010 00110

COMPUTER IQ

Machines that perform tasks that would be judged *intelligent*—if performed by a human—have been with us since antiquity. The Greek engineer, Hero of Alexandria, designed a hidden pneumatic system that would "miraculously" open the massive doors of a temple to allow entry to an arriving pilgrim when he kindled a fire in a small altar outside. This anticipates the modern supermarket's infrared-activated sliding door by many centuries.

A second example is Joseph Jacquard's of France 1805 automatic loom. Its complex behavior was guided by a series of stiff cards strung together in a chain. Prepunched holes in each card would block or permit the entry of various

**Patricia Churchland
and
Paul Churchland**

spring-loaded wooden pegs. The loom would weave a specific pattern into the emerging cloth as a result. Jacquard's loom could perform a variety of different weaving tasks, depending on the punchcards inserted. This loom is an early example of a programmable machine. The chain of punchcards was its program.

A third example, more impressive still, is the modern cruise missile, a small, unmanned jet aircraft containing a TV camera in its nose, an internal map of the territory to be traversed, and a set of instructions (the functional analog to Jacquard's punchcards) for matching its television images to its internal map in order to find its ultimate target. Like its predecessors, this "intelligent" machine works very well.

On the other hand, almost no one thinks that these machines display "real" intelligence. The mechanical or electronic reconstruction of some isolated human skill or behavior does not constitute real intelligence. And yet, many researchers do assert that, if all of the skills and all of the perceptual and behavioral capacities of a human were to be mechanically or electronically reconstructed in a unified physical system, then that machine would indeed possess genuine intelligence, no less than we do. The difference between humans, door openers, automatic looms, and cruise missiles—according to the modern consensus—is a difference in degree, not in kind. Human and animal intelligence, on this view, is just a very complex mixture of subsystems pieced together over millions of years of evolution. We may all agree that door openers and looms are too feeble to invite the honorific term, "intelligence," but in the resourceful cruise missile, we all see at least some

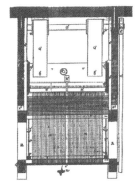

▲
THE JAQUARD MACHINE,
AS SHOWN IN
SCIENTIFIC AMERICAN
IN FEBRUARY 1846.

◄
DETAIL OF THE
PARTIALLY COMPLETED
DIFFERENCE ENGINE
NO.1, AN AUTOMATIC
CALCULATOR DESIGNED
BY CHARLES BABBAGE
IN THE 1820S.

elements of real intelligence, just as we see them in the behavior of an insect. More capable machines, it is said, will eventually earn the unqualified right to that term.

Research over the last forty years has lent strong support to this view of intelligence. Two broad approaches have dominated. The first exploits the modern digital computer and the second the lessons of modern neurobiology. They have at times been fierce adversaries in the early 1960s and again in the late 1980s, but they need not be mutually exclusive.

The most familiar is the classical or program-writing approach. The modern digital computer is, like Jacquard's loom, a *general-purpose* machine. Depending on the specific set of instructions, or program, we load into it, we can give it the capacity to respond, in any fashion we desire, to any of an indefinite range of possible inputs. So long as we can specify, in a finite set of instructions, the complex input-output patterns we desire, a digital computer can be made to display them. The mathematicians Alonzo Church (US) and Alan Turing (UK) had shown by 1950 that there are no specifiable input-output patterns (called computable functions by mathematicians) that lie beyond the theoretical capacity of a digital computer to embody, compute, and display in real behavior.

This mathematical result gave enormous encouragement to those who hoped to create machine intelligence by way of writing a suitable program—a program that, when run on the nearest available computer, would recreate the complex and elusive input-output function that presumably constitutes human intelligence. The Church/Turing result guaranteed that, on the modest assumption that human intelligence is indeed constituted by some elusive, but ultimately specifiable function, then a digital machine can be programmed to embody it. (The alternative to the "modest assumption" is that human intelligence is forever beyond rational comprehension. Few took this alternative seriously.)

In the early 1960s, programmers focused their efforts on isolated aspects of human intelligence, such as the ability to play chess, solve algebraic equations, and prove theorems in logic. Computer programs running on general-purpose machines are now very common; their uses include interactive video games, word processing, spread-sheet business planning, controlling the engine of one's car, landing giant airlines in any weather, making stock-market transactions, and even medical diagnosis. Such performances regularly exceed that of a human, often by a wide margin.

The programming approach has, however, been disappointing in simulating other tasks routinely performed by insects, birds, and mammals. Humans can quickly and easily do such things as recognize objects visually from almost any angle,

▲

EARLY IMAGINED

ROBOT DESIGN

pick out a familiar voice from a group of others, comprehend a conversation, distinguish an angry from a sad face, and walk smoothly over uneven ground— all of which have proven stoutly resistant to the programmer's art. In this "easy-for-us" domain, programs are typically clumsy, inflexible, and prone to common sense errors. The skeptical say that true intelligence is impossible for purely physical machines. A wiser assessment locates the problem in the digital-coding, serial-processing design of conventional computers. Such machines execute a program's instructions one operation at a time, one after the other, in serial fashion. They are fast— perhaps 100 million operations per second. But if the cognitive task at hand happens to require 100 billion operations, the machine will take 1,000 seconds, or more than 15 minutes, to execute it. A creature that cannot recognize predators faster than that is doomed to be somebody's lunch.

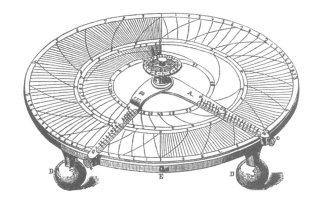

▲

Fortunately, the brains of living creatures display a very different computational design. The 100 billion neurons in a human brain each perform their individual computations simultaneously and in parallel, rather than one after another, in series. And they are connected to each other so that their collective behavior yields the complex input-output functions that a real-world creature needs. The result is that the parallel-computing brain computes its functions in small fractions of a second, rather than in the minutes, or hours, that a serial machine would need to accomplish the same task.

Computer designs modeled on the brain thus promise a way around the bottleneck of a serial processor. They are the focus of a new wave of research into machine intelligence, called neural network computing or parallel distributed processing. Beyond sheer speed, these networks display other features of interest, such as functional persistence despite physical damage, the flexibility to recognize things despite ambiguity and noise, and a sensitivity to complex analogies. Most importantly, they acquire their skills through extended learning from many examples, rather than from the guidance of a predesigned program. Like biological brains, they adapt themselves to experience. Artificial neural networks (ANNs) will never wholly displace the classical or programming approach; it is too endlessly useful for that. But ANNs have broken at least one logjam and launched our research in new directions. Although the IQ of machine intelligence is still low, the coming years will see it continue to rise

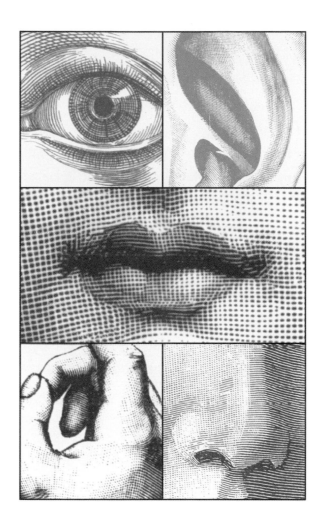

A CIRCUITOUS ROUTE

Our memories are vital to our identities. They tell us who we are and whom we love, what we do and how to do it, where we were and where to go, what has happened and what we imagine might happen. Memories are an essential feature of consciousness, but they also haunt our dreams.

Among all the members of earth's animal kingdom, humans have a unique capacity of imaging the future based on past experience. Predictions arise from the memories of countless images we sense, integrate, and record throughout our lives. With memory and predictions, we make choices that allow us to adapt and survive. These remembered images are of what we have seen, heard, touched, tasted or smelled, and are associated with feelings like love, fear, or anger—all in the service of satisfying primal needs, such as hunger or the avoidance of pain.

Daniel L. Alkon

While the ancients understood that our brains participate in creating memory, it was not until about four centuries ago that physical evidence began accumulating to support this hypothesis. The most powerful initial evidence came from the clinic where individuals with particular memory deficits were found to have lesions such as tumors in particular locations within the brain. French nineteenth century scholar P. Broca, for example, described a patient who lost his ability to speak, although he could hear and make the sounds of speech. A lesion in this patient's brain found at autopsy had apparently impaired memories necessary for vocalizing language.

Numerous correlations of pathologic lesions and cognitive deficits observed over the past two centuries have offered clues as to where memories for a variety of experiences might be stored. Actual functions of brain structures in acquiring, storing and recalling memories, however, are terribly difficult to discover. More sophisticated technologies and large numbers of experimental observations have clarified, but still not resolved, many of the key questions.

Pinpoint electrical stimulation of cells in certain brain regions, for example, has elicited memories in awake patients and certain behaviors in animals. Modern imaging techniques suggest that distinct regions of human brains become active at different times in the memory-storage process. However, most of our current understanding of how learning and memory actually occur among billions of neurons and their trillions of synaptic connections has emerged from studying animal brains.

What we have today is a fuzzy picture, only an impression of how our brains accomplish the task of recording experience. The brain's primary interest is in relationships. It does not record isolated bits of information, but rather the spatial and temporal relationships of bits within large constellations of bits. We remember a face, for example, by learning the spatial relationships of the face's features such as nose, mouth, eyes, forehead, chin, hair, and ears. We remember the word that names the face, for example, "Mother," by learning the temporal relationships between the sequence of sounds that accompany the letters of the word, as

well as the sequence of the visual images of the letters themselves, and the sequence of movements we must execute to make these sounds. We also learn the relationships between bit constellations. The bit constellation or image of the face is related in time to the sound image of the word "Mother."

In order to be learned, each of these images must pass through many successive stations within the nervous system. A visual image, for example, is first sensed by receptor cells in the eye and transmitted as electrical signals through the neuronal networks of the retina. Signals flow, in turn, from the retina to the optic nerve, the lateral geniculate, to the primary visual cortex and then to the hippocampus and different cortical areas including the parietal cortex and the frontal cortex.

The signals from the eye, a fairly small structure, then, fan out, spreading through and across much of the brain. As the signals fan out, information is processed, and at many brain stations local networks are activated and undergo temporary and eventually permanent changes in order to store memories of the image.

There is apparently a long sequence of cellular changes extending from brief to very long-lasting time domains. Within local synaptic networks of cells, the image's electrical signals release chemical messengers, or transmitters, between neurons. Within neurons, neurotransmitters then trigger molecular activities that ultimately change the strength of the synaptic signals and, probably, the actual shape of the synaptic connections between neurons.

Learning, then, appears to involve subtle but significant rewiring within local networks that are distributed throughout many brain regions. Although our understanding is still primitive, it is already generating new approaches to diseases that affect memory, for example, dementia, such as Alzheimer's disease. We are also beginning to extract principles of how the brain's networks store and access memory. These principles can be expressed mathematically, and can be manipulated with computer simulations that show a promising capacity to learn images that they "sense" through a video monitor.

BOUND FOR GLORY

From the beginning, chemistry was the craft of making substances change, or watching their spontaneous transformations. Ice turned into water, water could be made to boil. Grape juice or sugar cane mash turned into alcohol, and if one didn't intervene, it turned again, into vinegar. A colorless fluid from the gland of a Mediterranean sea snail, when exposed to air and sunlight, turned to yellow, then to green, and finally to a purple that could dye fast a skein of wool.

Roald Hoffmann

Today, chemistry is the science of molecules and their transformations. Over a period roughly coinciding with the history of *Scientific American*, the craft became a science, and instead of studying substances, chemists now think of molecules.

Molecules are made up of atoms. There are about 90 natural elements or kinds of atoms, 15 or so radioactive, humanly created ones. Some matter is truly atomic in its composition (helium or argon gas); some is made all of one atom, but with the atoms linked up in some simple or complex way (the iron atoms in iron metal; carbon in graphite or diamond).

But what a dull world it would be if there were only 105 things in it! Any square foot of this beautiful earth shows a far greater richness. The world is made of molecules—sugar, aspirin, DNA, bronze, hemoglobin—persistent groupings of atoms with reproducible colors, chemical properties, toxicity, which are a consequence not only of the identity of their atomic components, but also of the way those atoms are connected to each other.

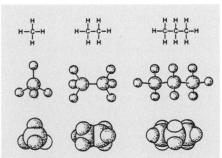

That connection between atoms is called a bond. It's not a random coupling; there are rules to this cross between a donnybrook and a love affair. So carbon typically binds to four others and hydrogen forms a liaison (indeed, that's the French word for bond) with one. And then the game is on between the two, for one does not have CH (at least not much of it; that wouldn't satisfy carbon's constrained lust for bonding, and when CH is found it is a most reactive unstable species) but CH_4, methane. One can also form carbon-carbon bonds, and the constructive game begins in earnest with the hydrocarbon series: methane, ethane, propane, etc. The chain builds; its approach to infinity is that ubiquitous polymer, the most important plastic of our times, polyethylene.

Individual molecules are really tiny. Consider the sugar (glucose) molecule drawn to scale here. Just as a map might have a scale of 1:250,000, meaning that an inch on the map is 250,000 inches in reality, so we could say that these drawings are roughly 174,000,000:1, 1.74 inches on this page corresponding to 0.00000001 inches in the molecule. A single glucose molecule is much too small to be seen with the eye, or even with the very best optical microscope.

▶

A "BALL-AND-STICK"
MODEL OF A GLUCOSE
MOLECULE

Our knowledge of the atoms in a given molecule, of how they are connected to each other, and even of the molecule's three-dimensional shape, is largely indirect. We use various instruments, perturbing the molecules (often by light) and measure their responses. The ingenious way analytical chemistry knows without seeing is an incredible achievement of this century.

In the gas or seething liquid, an individual molecule is buffeted by collisions with other molecules. It travels, gives, deforms, yet remains a molecule with a

Look-alikes

Most chemical compounds look alike— 90 percent are white crystalline solids. Yet some are beneficial, some harmful. In the realm of molecular differences, a particularly subtle but important one is chirality, or handedness. Some molecules exist in distinct mirror-image forms, related to each other as a left hand is to a right. Many, but not all, of the properties of such mirror-image molecules are the same— they have identical melting points, colors, etc. But some properties differ, often critically. This is, for instance, true of their interaction with other-handed molecules. Think of left feet meeting up with left or right shoes. So the enantiomers (for that is the name for the distinctly handed forms of a chiral molecule) *may have drastically different biological properties. One may taste sweet, its mirror form tasteless. The mirror-image form of morphine is a much less potent pain reliever.*

The illustration here shows d- and l-carvone. d-carvone can be isolated from caraway and dill seed, l-carvone from spearmint. And they are responsible for a good part of the taste and odor of these plants—they smell like caraway or spearmint—whether they are natural extracts or are made in the laboratory.

Our proteins, for instance the human smell and pain receptors, are like complex gloves; not always, but very often, they respond differently to handed molecules.

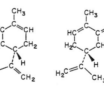

d- AND l-CARVONE

structure. The structural perspective is a fruitful world view, for we are builders from the beginning. You can build beautiful simple things, such as the inorganic $B_{12}H_{12}^{2-}$ ion, an icosahedron of borons, with twelve hydrogens pointing out.

Or you can build more complicated things. Below is a new, effective, immunosuppressant, FK-506.

Once you learn how to build it, you gain the power to fiddle with the design. If you can find out how this immunosuppressant fits the site that it binds to in a person's body, perhaps you can change a piece of the drug that seems to produce unwanted side-effects.

Part of the definition of chemistry has survived from medieval times to today: chemistry is change. While the atoms in a molecule persist in their association with each other, the input of energy—heat, light, electricity—can induce a change. From collisions on that busy dance floor inside a flask emerge regroupings, new associations of atoms, new molecules.

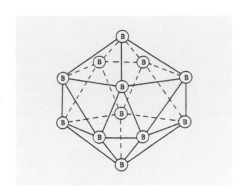

◀

A MODEL OF AN

ICOSAHEDRAL

$B_{12}H_{12}^{2-}$ ION

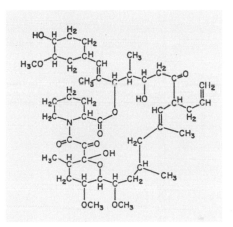

◀

THE

IMMUNOSUPPRESSANT

FK-506

Molecular dynamics, the study of molecules in motion or in chemical reactions, is one of the most exciting fields of modern chemistry. How to catch the fleeting molecules that come and go in a flame? What happens, in molecular detail, on the surface of a seemingly magic catalyst that reduces unwanted pollutants in the exhaust of a car (and what could replace the expensive rhodium in that catalyst)? Will a touted chlorofluorocarbon substitute in fact not harm the ozone layer?

Such questions are crucial to the vast chemical industry. Searching for the answers satisfies a basic human urge behind all science—curiosity. And the answers, which must be given in terms of how molecules interact with other molecules, will be essential for the comfort of people, and for the survival of the planet.

THE WAR MACHINE

1941

THE FIRST JET airplane is flown.

THE MANHATTAN PROJECT—the code name for the U.S. government's top-secret initiative to develop the atomic bomb—is begun by executive order of President Roosevelt on December 6. On December 7 Japan attacks Pearl Harbor, prompting the United States to declare war on Japan and moving Germany, as Japan's ally, to declare war on the United States.

42

THE FIRST SELF-SUSTAINING NUCLEAR chain reaction is created on a squash court at the University of Chicago by a team headed by Italian physicist Enrico Fermi.

43

OCEANOGRAPHER AND ACTIVE MEMBER of the French resistance Jacques-Yves Cousteau invents the first lightweight, self-contained underwater breathing apparatus or scuba device.

WHILE MODIFYING LYSERGIC ACID to form its diethylamide compound, Swiss chemist Albert Hoffman mistakenly absorbs some of the substance and begins to hallucinate. Intrigued, he deliberately ingests more. A generation will pass before Hoffman's accidental discovery becomes popularly known by its abbreviated name: LSD.

44

U.S. BACTERIOLOGIST OSWALD AVERY and his associates strip away the protective protein capsule surrounding killed pneumococcus bacteria and add what remains to a living nonencapsulated strain. When the bacterial descendents of this union inherit the missing capsule, Avery's team becomes the first to show that the genetic instructions to build it could only have come from the encapsulated bacteria's deoxyribonucleic acid, or DNA. The field of genetics is revolutionized.

OSWALD AVERY

45

GERMANY surrenders on May 7.

MANHATTAN PROJECT SCIENTISTS detonate the first nuclear fission bomb in the New Mexico desert on July 16. Although they expect an explosive force equivalent to 5,000 tons of TNT, the devastating result is four times greater. Just three weeks later, the B-29 bomber Enola Gay drops the first atomic bomb over Hiroshima, Japan, leveling 90 percent of the city and resulting in 130,000 people killed, injured, or missing. A few days later a more powerful plutonium-based bomb is exploded over Nagasaki leaving 75,000 people dead or wounded and devastating more than one-third of the city. Japan surrenders on Aug. 14, ending World War II.

46

THANKS TO HIGH-ALTITUDE maneuvers during the war, pilots begin to recognize a permanent band of westerly winds at the boundary between the troposphere and stratosphere. Hundreds of miles wide, several miles deep, and with speeds of up to 300 miles per hour, the bands come to be known as jet streams and are found to have a profound effect on earth's weather.

U.S. CHEMIST WILLARD LIBBY discovers radiocarbon dating, which quickly becomes an invaluable tool for accurately estimating the age of archaeological artifacts.

WITH THE SOVIET UNION DOMINATING Eastern Europe, an iron curtain falls between East and West. The cold war begins.

47

A GROUP OF IN-VESTORS headed by Gerard Piel, Dennis Flanagan and Donald Miller purchase and re-launch *Scientific American.*

48

RUSSIAN-BORN U.S. PHYSICIST George Gamow develops Belgian cosmologist George Lemaître's earlier idea that an expanding universe must once have been joined. Gamow terms the universe-creating explosion the big bang and theorizes how the chemical elements were formed in its aftermath. He predicts that the energy released by the explosion can still be seen as background radiation bathing all parts of the sky.

SNAPPING ATOMS TOGETHER

Manufactured products are made from atoms. The properties of those products depend on how those atoms are arranged. The graphite in a pencil lead is made from carbon atoms arranged in hexagonal sheets. Rearranged, those carbon atoms will make a diamond. Sand is made primarily of silicon and oxygen atoms. Rearranged, the same atoms (with traces of other elements) make the electronic chips that are the hearts of our television sets and computers.

Ralph C. Merkle
Chipping flint was an early means of removing untold billions of unwanted atoms from an arrowhead, while beating on hot steel with a hammer pushed other billions of atoms into the shape of a sword. Modern manufacturing methods are more precise. Lithography (literally "stone writing") lets us make the intricate pattern of wires and logic gates in a modern computer on a thin wafer of silicon. Millions of transistors can already be put on a few square centimeters of a chip, and it is expected that by the turn of the century, we will be able to put a billion transistors on a chip.

In lithography, a pattern of light and dark is projected onto the surface of a silicon wafer, just as a camera projects such a pattern onto film. An optically sensitive coating on the wafer is then developed and selectively washed away. Different types of molecules can be sprayed on the wafer, but will only bond to the actual silicon surface where the coating has been removed. The remaining coating is washed away, leaving only the bare silicon with a complex two-dimensional pattern of the desired material on it. This process is repeated many times using different patterns and different types of material to make working transistors and wires.

As the lithographic picture projected onto the silicon wafer is made sharper—and the associated etching and deposition steps are made more precise—the size of individual transistors can be made smaller and more will fit on the wafer. Over the last several decades, refinement has resulted in ever-more, ever-cheaper transistors. Today's lithography can make features that measure less than a micron (one millionth of a meter). Improvements in lithography are expected to continue for another decade, but fundamental limits will eventually bring progress to a halt. Cost is also becoming more important. Chip-making factories already cost $1 billion and are expected to cost $10 billion in less than a decade. To continue revolutionizing computer hardware, we will have to introduce new, post-lithographic manufacturing methods that are both more precise and less costly.

◄

CARBON'S AGILE ATOMS CAN FORM A SUBSTANCE AS HARD AS A DIAMOND OR AS SOFT AS GRAPHITE.

With today's manufacturing methods (even lithography) we are only able to move atoms in statistical herds; this process is like building something out of Lego blocks while wearing boxing gloves. You could scoop up lots of blocks and push them into heaps, but you couldn't snap the blocks together the way you want. Many scientists believe that nanotechnology (also called molecular manufacturing, when the broader term is ambiguous) will allow us to take off our boxing gloves and build products in which essentially every atom is in the right place. This new technology will let us achieve the finest possible precision and will vastly improve almost all manufactured products. Computer technology will be the most obvious beneficiary. As the size of transistors in a computer shrinks, the position of each atom becomes more and more important. To reach the limits of miniaturization, we will have to put each atom in a precise position dictated by a complex design, and we will have to do this with billions of atoms.

On the macroscopic scale, the idea of manufacturing products by taking the parts in our hands and putting them together is commonplace. On the molecular scale, the concept of holding and positioning molecular parts and snapping them together into a product is new. Chemists have been synthesizing molecules for centuries, but without any "molecular hands" to help them. Molecules are usually in solution (in a test tube), bouncing at random under the influence of thermal noise. They are designed to assemble spontaneously when they are mixed together. If we wanted to build a radio this way, we would first take all its parts and put them

Feynman's Dream

In a remarkable talk given in 1959, Nobel Prize-winning physicist Richard P. Feynman talked about miniaturization and its ultimate limits. He said, "The principles of physics, as far as I can see, do not speak against the possibility of maneuvering things atom by atom." He asked then the question that many scientists are asking today: "What would happen if we could arrange the atoms one by one the way we want them (within reason, of course; you can't put them so that they are chemically unstable, for example)."

And he gave a prophetic answer: "The problems of chemistry and biology can be greatly helped if our ability to see what we are doing, and to do things on an atomic level, is ultimately developed—a development which I think cannot be avoided."

into a bag, shake it, and the parts would spontaneously assemble into a radio— a very hard way to build a radio. The more parts there are that need assembling, and the more complex the pattern into which we wish to assemble them, the harder it is to take this approach. That it is possible to position individual atoms and molecules has already been demonstrated experimentally in simple systems. It will take many years and a great deal of work before we can use this method to build more complex structures, such as transistors and intricate wiring needed for a molecular computer. But many scientists think this new era is inevitable.

The ability to make, at great expense, a few very small machines by building them molecule by molecule will not, by itself, let us make large computers or other products economically. We know, however, that nature makes large structures such as trees by using small programmable self-replicating machines called cells; here is a procedure we can try to imitate.

John Von Neumann, a Hungarian mathematician working at Princeton proposed in the late 1940s a self-replicating system that is essentially a computer connected to a robotic arm. The computer directs the arm to pick up and assemble parts into a second computer and a second arm, and so the system can copy itself. K. Eric Drexler (US) has proposed the "assembler," which uses a molecular computer to direct a molecular robotic arm which snaps together individual molecules into almost any desired structure, including another assembler. Self-replicating assemblers would be inexpensive, and groups of them could later be programmed to make almost any desired product, just as cells can be programmed by changing their DNA to make trees, wheat, corn, etc.

To take but one example of the profound effect this new ability would have on our lives, consider that most disease is caused by damage at the molecular and cellular level. Today, we don't have the tools to intervene at that level. With nanotechnology, we could make a fleet of tiny robot surgeons with molecular tools, guided by molecular computers that could attack sickness at its roots. This technology would revolutionize the practice of medicine.

During the past 150 years we have seen our manufacturing technologies steadily improve, becoming ever more precise and ever more flexible. We can now see on the horizon the ultimate goal: manufacturing technology able to arrange inexpensively the fundamental building blocks of matter in almost any arrangement consistent with physical law. Revolutionary advances in computers and medicine are only two of the many benefits this should bring. Achieving this goal will require the cooperative work of many people from many disciplines over many years.

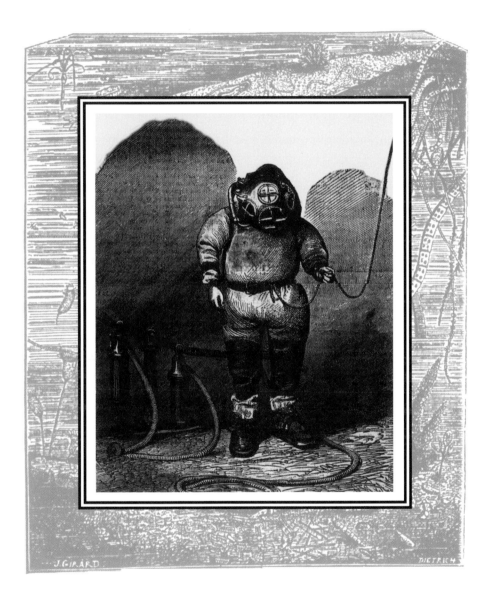

THE FRONTIER BELOW

As the twenty-first century nears, some believe that the earth is now fully explored, and turn their gaze skyward seeking new frontiers. Ocean scientists, however, know

Sylvia A. Earle differently—the greatest era of exploration on this planet has only just begun. We have traversed the surface of the sea for millenia, but only recently have we gained effective access to the ocean depths. While the sea is the central to our planet's life-support system, the crucible that forges climate, weather, and planetary chemistry, and home to most of life on earth, little is known of that vast three-dimensional realm nor of the diverse creatures that dwell there.

One hundred fifty years ago, diving bells enabled a few intrepid explorers to make brief descents underwater to more than 30 meters and, holding their breath, divers sometimes went to twice that depth—but only for a minute or so at a time. Even earlier, by 1800, several experimental diving suits had been designed and tested; a clever Englishman named John Lethbridge had invented a diving barrel equipped with a small glass viewport, and Robert Fulton had successfully launched the Nautilus, the first working submarine.

Many other ocean "firsts" followed in the 1800s. The first practical commercial diving systems were developed by Augustus Siebe in 1819, the first closed breathing system for divers was introduced by Benoit Rouquayrol and Lt. Auguste Denayrouze in 1865; and the expedition of HMS *Challenger*, the first global oceanographic voyage, set off in 1872, heralding the birth of the science of oceanography.

The instructions to *Challenger's* captain read, "...you have been abundantly supplied with all the instruments and apparatus which modern science and practical experience have been able to suggest and devise...you have a wide field and virgin ground before you."

Scientists aboard *Challenger* used nets, dredges, water samplers and other devices lowered from the deck of the ship. Although metal helmets and canvas suits supplied with compressed air had been in use for half a century, no such equipment was taken to sea for the four-year exploratory voyage. Masks and flippers did not yet exist, and the development of the *self-contained underwater breathing apparatus*—scuba—was still many years away.

Those who set out to explore the sea a century ago were like visitors trying to explore our cities, but able only to do so by flying 1000 meters above, blindly lowering instruments, dragging hooks and nets through the streets and buildings

◄

SUBMARINE DIVERS' EQUIPMENT WEIGHED MORE THAN 140 POUNDS IN 1873, SHOWN IN THIS ENGRAVING FROM *SCIENTIFIC AMERICAN*.

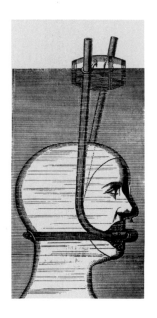

below, and dredging up a random mix of anything they encountered—bushes, fragments of cement, puzzled pedestrians. With such haphazard methods, how much real knowledge could possibly be gained about who we are, how we live and what we do? How much has been learned about the nature of the sea, especially the deep sea, using such methods? Actually, a surprising amount, at least in terms of the subsurface terrain, and the broad patterns of water movement and chemistry.

Until the 1950s, however, the most conspicuous geological feature of earth remained hidden. This was a 60,000-kilometer mountain range running along the middle of the Atlantic, Pacific and Indian oceans. Much of our basic knowledge concerning the earth's basic geology and geophysics—including the now-familiar concept of plate tectonics—became accessible once we were able to probe beneath the ocean's surface with sound and more effective submersibles.

Ventures into the deep were pioneered in the 1930s by zoologist William Beebe and engineer Otis Barton using a tethered submersible, the Bathysphere, designed and built by Barton for $11,000. Inside the hollow steel ball, Beebe and Barton were the first to glimpse deep-sea creatures in their own habitats, half a mile beneath the ocean's surface. At about the same time, a new kind of diving suit was being perfected by Englishman Joseph Peress, a system called *Jim* after Jim Jarrett, the first person willing to try it. *Jim* looks and functions much like an astronaut's space suit. Inside, pressure is maintained at one atmosphere, and air is supplied through a rebreathing system that adds oxygen as needed and chemically removes exhaled carbon dioxide.

A major advance in ocean exploration came in 1943 when Jacques Cousteau and engineer Emile Gagnan introduced a device that freed divers from underwater tethers—the aqualung. At first considered daring and dangerous, it enabled millions of people to explore the ocean depths. Professional scientists and recreational divers alike have transformed our understanding of the ocean—at least the uppermost 50 meters of it. Below, access has remained elusive, and few have glimpsed the ocean below a few hundred meters.

Only two men, U.S. Navy Lieutenant Don Walsh and Swiss engineer Jacques Piccard, have descended to—and returned from—the deepest part of the sea, 11,000 meters down. In 1960, the bathyscaphe Trieste transported the two explorers to the bottom of the Marianas Trench near the Philippine Islands, an accomplishment as significant in its way as the first visit to the moon by Neil Armstrong and Buzz Aldrin nearly a decade later. Walsh and Piccard were able to answer what to many was the ultimate question—whether or not life existed in the deepest sea. They were greeted by a flounder-like fish living in a dark, cold atmosphere where the pressure is 16,000 pounds per square inch. The discovery gave rise to numerous new questions about how life forms survive in what is to us a hostile deep-sea environment.

More questions and a few answers have come from more recent deep-sea research by scientists using submersibles and a growing number of remotely operated and towed vehicles equipped with cameras and various other instruments. Among the most active of the small number of manned submersibles used for ocean research is *Alvin*, launched in 1964 by the Woods Hole Oceanographic Institution and in continuous operation ever since. In 1977, during dives to 2000 meters along the Galapagos Islands, scientists had their first glimpse of communities of animals whose existence revolutionized theories about life in the oceans. Giant tube worms, white crabs, strange anemones, large bivalve mollusks and numerous other previously unseen creatures were found living on nutrients derived from chemosynthetic sulphur bacteria. Those bacteria, in turn, were thriving on hydrogen sulphide and minerals pouring from hot deep-sea springs associated with volcanic vents in the sea floor.

Expeditions using other submersibles in the past two decades have brought new insight and many surprises concerning the abundance and diversity of life in the sea as well as new information about geological processes. Meanwhile, the grand overview made possible by space flight has yielded a wealth of data concerning sea surface temperature, currents, waves, and even the distribution of certain kinds of marine organisms. Instruments and cameras mounted on satellites provide real-time information about basic earth processes that scientists aboard the ocean-going *Challenger* could only dream about. Today, technology has virtually transformed deep-sea exploration—with sonar, lasers, electronic communications, computers, biotechnology, submersibles, and underwater vehicles piloted from thousands of kilometers away.

In an era when space probes travel beyond the solar system and plans are underway for humans to visit our sister planet Mars millions of kilometers away, we must realize that it is not presently possible to travel to the deepest sea, a mere 11 kilometers straight down. Not since 1960 have we seen the ocean's furthest depths, and not until 1994 did a tethered underwater robot, the Japanese vehicle *Kaiko*, return there. Presently, only five manned submersibles and about as many towed and autonomous unmanned vehicles exist that can transport observers even halfway to the bottom of the ocean. As the present millenium comes to a close, less than one tenth of one per cent of the ocean below 50 meters has been explored.

▲

"DEVIL FISH" SHOWN
IN *SCIENTIFIC
AMERICAN* IN 1873

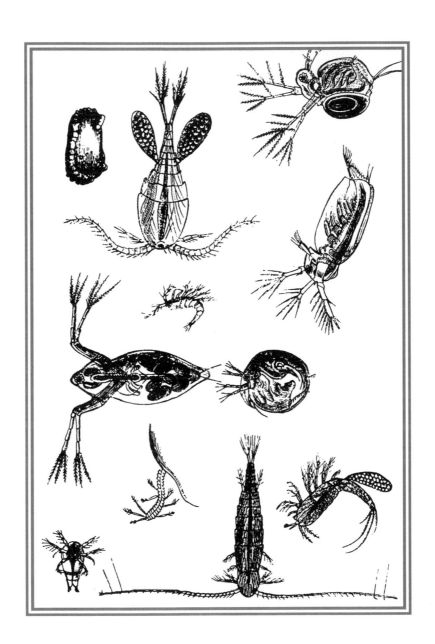

THE EMERGENCE OF LIFE

More than 125 years ago, in his *Fragments of Science for Unscientific People*, the British physicist John Tyndall wrote: "Darwin placed at the root of life a primordial germ, from which he conceived that the amazing richness and variety of life now upon the earth's surface might have been deduced. If this hypothesis were true, it would not be final. The human imagination would infallibly look beyond the germ and however hopeless the attempt, would enquire into the history of its genesis... A desire immediately arises to connect the present life of our planet with the past. We wish to know something of our remotest ancestry... Does life belong to what we call matter, or is it an independent principle, inserted into matter at some suitable epoch, when the physical conditions became such as to permit the development of life?"

Cyril Ponnamperuma

Indeed, Darwin himself had wondered about what may have transpired before the appearance of that first germ from which he believed that all biology had evolved. At a time when there was much dispute over the origin of species, little or no attempt was made to further understand what may have happened at the dawn of life. Yet, in a very telling passage, he wrote to his friend, Hooker, "But if, (and oh what a big if!) we could conceive in some warm little pond with all sorts of ammonia and phosphoric salts, light, heat, electricity, etc., present that a protein compound was chemically formed ready to undergo still more complex changes." Here is the seminal idea that has motivated scientists during the last one hundred and fifty years in their search for the origins of life.

Long before the theory of biological evolution revolutionized scientific thinking, spontaneous generation had been accepted as nature's fundamental lifegiving process. One had only to accept the evidence of the senses, thought the ancients: worms from mud, maggots from decaying meat, and mice from old linen. The Egyptians believed this. Aristotle taught the same doctrine in his Metaphysics. Newton, Harvey, Descartes, Van Helmont all accepted this without serious question. Then Louis Pasteur, with his rigorous experimentation, dealt a mortal blow to the time-honored concept. The story of Pasteur is often told to beginning students in biology as an example of triumph of reason over

◄

LONG BEFORE THE THEORY OF BIOLOGICAL EVOLUTION REVOLUTIONIZED SCIENTIFIC THINKING, SPONTANEOUS GENERATION HAD BEEN ACCEPTED AS NATURE'S FUNDAMENTAL LIFEGIVING PROCESS.

mysticism, but today we have redefined spontaneous generation—in the molecular sense—as chemical evolution.

Alexander Ivanovich Oparin, the Russian biochemist, in 1924 was the first scientist in the twentieth century to postulate in scientifically defensible terms that chemical evolution preceded biological evolution. In 1928, independently of Oparin, the British biologist J.B.S. Haldane speculated on the early conditions suitable for the emergence of terrestrial life. Haldane suggested that sunlight, acting on the earth's early atmosphere, produced a "hot dilute soup," giving rise to the popular notion of the "primordial soup." Laboratory testing of the Oparin-Haldane hypothesis began in 1951, when Melvin Calvin used the Berkeley cyclotron to produce some of the molecules basic to life. Another much-cited experiment was performed by Harold Urey and Stanley Miller in 1953. Here, ammonia, hydrogen and methane were subjected to an electric discharge which simulated primeval lightning. Four of the amino acids commonly found in protein were synthesized in this experiment. A long era of experimentation followed. To date, laboratory work has clearly demonstrated that most of the molecules necessary for life could have been formed by the action of natural forces on the primitive atmosphere. Such processes could also occur in the atmospheres of other planetary bodies. The comet Shoemaker-Levy 9, crashing into the atmosphere of Jupiter in July of 1994, gave us a grandstand view of such a phenomenon.

Modern biochemistry has revealed the importance of the nucleic acids (DNA and RNA) and the proteins. While DNA and RNA provide the blueprint of life, the proteins form the framework of every living organism. These molecules consist of 28 basic components—the alphabet of life!

At the Bottom of the Sea

Is new life arising now? In our studies of the origin of life we had assumed that conditions necessary for the emergence of life had disappeared from the face of the earth. We attempted, therefore, to recreate them in the laboratory. The sub-marine research vessel, the Alvin, has discovered active volcanoes and hy-

drothermal vents at the bottom of the ocean, whose environments closely replicate conditions on the earth's surface four billion years ago. Microbes have been discovered in these areas of extreme temperatures and pressures. Perhaps life begins at any time, wherever suitable conditions exist.

Did chemical evolution really happen? To answer this question we must go back to the very birth of our planet. By a painstaking examination of the ancient sediments of South Africa and Australia, Elso Barghoorn of Harvard and his colleagues William Schopf and Andrew Knoll have established the abundance of microbial life on earth as far back as 3.5 billion years ago.

Since we have not been able to find any trace of early chemical evolution of earth, we have sought such evidence in extraterrestrial material from the moon and from meteorites. Analysis of moon rocks brought back by the Apollo astronauts showed little evidence of any chemical evolution on the moon. The high vacuum and intense solar radiation on the lunar surface had destroyed any trace of organic matter. However, while the lunar analysis was inconclusive, the study of meteorites was more fruitful. A meteorite that fell at Murchison in Australia in 1969 gave us the first convincing evidence of the presence of extraterrestrial amino acids in the early solar system. These findings have been extended to a larger cache of meteorites found in the Antarctic in the last twenty years. Moving beyond our own solar system, radioastronomers studying the interstellar medium have detected a vast array of organic molecules, including some of the precursors of life. Organic chemistry appears to be common throughout the cosmos; the building blocks of life are therefore readily available anywhere in the universe.

Is there life beyond earth? Modern astronomy has shown that our sun is just one among billions of stars in the cosmos. Planets are therefore plentiful in the universe. In July 1976, to mark the two-hundredth anniversary of the birth of the United States, two spacecrafts, Viking I and Viking II, were sent on a mission. These spacecrafts carried a number of robotic devices capable of performing specific life-detection experiments. Although the biology experiments produced tentative results, the chemical experiments clearly indicated that there was no organic matter on the Martian surface. And since the search for life on Mars was predicated on the assumption that life is based on organic matter, it was concluded that the Viking mission provided no evidence for life on Mars.

Radio communication is yet another tool with which to search for life elsewhere in the universe. A program entitled SETI (Search for ExtraTerrestrial Intelligence), launched by NASA and maintained with support from international organizations, focuses on the possibility of "galactic eavesdropping." Perhaps our cosmic neighbors have been monitoring our progress since the days of Marconi, and are even now tuning in to "Radio Earth!" Our odyssey began with the study of the emergence of life on earth, and has inspired us, like Ulysses, to explore "what lies beyond the sun."

BECOMING HUMAN

Anthropologists have long been aware of the close anatomical, behavioral and genetic similarities between humans and the African apes. According to Charles Darwin, these similarities must be the result of our having shared a common ancestor in the distant past. Sometime between five and eleven million years ago in Africa, the common ancestor split into two lineages; one evolved into modern African apes and the other into humans.

Donald C. Johanson

A definitive fossil record for human evolution begins about 4.4 million years ago, with the appearance of the oldest known hominid (humans and our direct ancestors) species, called Australopithecus ramidus. This recently discovered species (named in 1994) is known from fossil fragments found at Aramis, Ethiopia. It has the most apelike teeth of any hominid found to date. The oldest well-known hominid species is Australopithecus afareniss, a descendant of A. ramidus which appeared about 3.9 million years ago. Hundreds of specimens have been recovered from several sites in the Great Rift Valley, from northern Tanzania to northeastern Ethiopia. This species' most important representative is a 3.2 million-year old partial skeleton, affectionately called Lucy, found at the site of Hadar in Ethiopia.

A. afarensis possessed numerous apelike features, including its teeth, projecting face, and small-capacity skull. As is the case with some African apes, males were much larger than females. But unlike the quadrupedal apes, the early hominids locomoted bipedally. Some interpret the apelike features seen in the arms and legs as indicating climbing activities in this ancestor. However, the three and a half million-year old hominid footprints left in a volcanic ash, at Laetoli, Tanzania, are virtually identical to those made by modern humans in soft beach sand.

Between three and four million years ago, the anatomy of A. afarensis remained relatively unchanged. However, beginning around three million years ago, the species gave rise to a number of distinct lineages; the exact number is still being debated. The diversification in hominid types may have occurred in response to climatic change; a cooling and drying trend. In one lineage, reduction in size of the jaws, teeth and supporting bone structures for the attachment of chewing muscles was accompanied by steady increase in brain size. This line of evolution is usually labeled the Homo lineage and may have first emerged in eastern Africa. It is also thought to be associated with the oldest stone tools, which appear about two and a half million years ago in Ethiopia.

◄

THIS ANCESTRAL HUMAN SKELETON WAS FOUND IN MENTONE, A VILLAGE ON THE SOUTH COAST OF FRANCE, IN 1873.

In southern Africa about 2.8 million years ago, another evolutionary trend was underway. Named A. africanus, this species developed large jaws with expanded chewing teeth. Continued specialization in this lineage gave rise to a form called A. robustus, which had quite large chewing teeth and very heavy jaws; larger individuals sported a crest on top of their skulls to which were attached chewing muscles. A. robustus was a specialized hominid and subsisted on a diet of particularly tough vegetables.

In eastern Africa, a parallel development was occurring in hominid evolution. Specimens assigned to A. aethiopicus date to about two and a half million years and, like A. robustus, show a number of heavy-grinding adaptations, such as very

large back teeth and a prominent, blade-like crest atop the skull. The strongly projecting face is reminiscent of A. afarensis, its probable ancestor. The exact evolutionary relationships between A. africanus, A. robustus and A. aethiopicus are debated continuously. However, A. aethiopicus gave rise to an ultra-vegetarian species, known as A. boisei in eastern Africa. The vegetarians were long-lived and widespread but finally met their extinction about one million years ago.

The larger-brained species, Homo habilis, dates from approximately 2.3 to about 1.6 million years ago, and was first identified at Olduvai Gorge in 1964. Further specimens have been found showing additional anatomical variation, suggesting there may have been more than one species of Homo during this time. These beings show brain expansion and a reduction in tooth and jaw size. They most likely made the simple stone flakes and pebbles found in association with cut and broken animals bones. Augmenting their predominately vegetarian diet with scavenged meat, H. habilis was beginning to live a different life style.

The oldest evidence for Homo erectus, a descendent of H. habilis, dated to 1.7 million years ago, is from Lake Turkana in Kenya. This species was tall and slender with long upper and lower limbs—a physique suited for the tropics. With even smaller chewing teeth, a forward projecting humanlike nose and a larger brain, H. erectus was more like modern humans than any of its forerunners. Shortly after making its appearance in Africa, H. erectus migrated into Eurasia.

The ability of H. erectus to leave the tropics and begin living in cooler and more temperate climates was perhaps caused by increased hunting or scavenging associated with more sophisticated stone tools. Around one million years ago, H. erectus probably also began to control fire.

H. erectus lasted until about 250,000 years ago, but it is unclear if all specimens belong to a single species. "Archaic" Homo sapiens, known from Europe and Africa, combined advanced features such as a more rounded and expanded braincase with older H. erectus features, like large brow ridges. They were probably the predecessors to Neandertals and modern humans.

One group of "Archaic" Homo sapiens, isolated in southern and western Europe by glaciers, developed a distinctive anatomy in response to the colder climate. Referred to as Homo neanderthalensis, they had long, low skulls, very large sinuses and short, squat heat-conserving body shapes. They buried their dead, albeit rather unceremoniously, made clothes, hunted to some extent, and possessed a wide range of flake tools, but they did not create art.

Anatomically modern humans first appear in Africa just prior to 100,000 years ago, but it was not until approximately 50,000 years ago that Homo sapiens became behaviorally modern as well. At that time, they developed a more sophisticated culture with a quite diverse tool kit that was based primarily on blade tools rather than flake tools. Homo sapiens carved bone, antler and ivory, and buried their dead with elaborate grave goods. Their ability to depict the world around them is seen in their art, particularly in the paintings which have been found decorating some cave walls like Lascaux in France. Homo sapiens' becoming fully modern may reflect an important series of developments in the neurology of the brain which led to a heightened level of consciousness, as well as modern language ability.

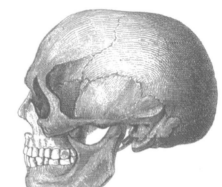

Shortly after Homo sapiens moved into Europe, about 32,000 years ago, Homo neanderthalensis vanished. Researchers are now debating two competing explanations to account for the origin and dispersal of modern humans around the globe. One model, "regional continuity" has modern humans arising regionally from isolated "archaic" populations. The other, known as "out of Africa," suggests that Homo sapiens evolved in Africa and then spread outward. The latter model is supported by the discovery in Africa of the earliest Homo sapiens and is consistent with the extensive similarities seen between all living human populations, which exhibit only minor differences in genetic makeup. Variations in external features like skin, eye and hair color, facial shape and so on are of minor significance. An anthropologist studying skulls assembled from around the contemporary world would easily classify them as a single species—Homo sapiens.

1950

NORTH KOREA INVADES South Korea, beginning the Korean War. United States and Chinese forces enter the conflict.

▲
NEAR-INFRARED IMAGE OF THE MILKY WAY

52

FOR HIS DOCTORAL DISSERTATION, U.S. chemist Stanley Miller recreates the primitive earth in the laboratory of his advisor, Harold Urey. His simulation includes heat to foster evaporation, a cooling tube to allow condensation and rainfall, and sparks to generate lightning energy. After a week, Miller finds his tube full of organic compounds and simple amino acids, the basic building blocks of protein and hence all life. A year later a Gallop poll finds that 22 percent of the respondents do not exclude the possibility of creating life in a test tube.

51

RADIO TELESCOPES REVEAL information about the universe not detectable by optical means. Radio waves characteristic of the spiral arms of galaxies are found coming from our own galaxy. Astronomer William Morgan discovers that the Milky Way galaxy is a spiral galaxy, and that our sun is located in one of its outer arms.

A COW IS SUCCESSFULLY IMPREGNATED with frozen semen, and the first sex-change operation is performed on George Jorgenson—who becomes known to the world as Christine.

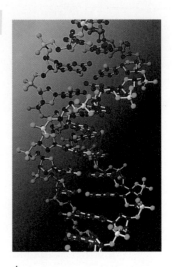

▲
COMPUTER-GENERATED MODEL OF A STRAND OF B-DNA

53

U.S. BIOLOGIST JAMES WATSON and British biologist Francis Crick announce to patrons at their favorite pub that they have discovered "the secret of life." Their model of deoxyribonucleic acid, or DNA, accurately describes the molecule that transmits hereditary characteristics from parent to offspring in living organisms. Watson and Crick, and British biophysicist Maurice Wilkins, share a Nobel Prize for their discovery in 1962.

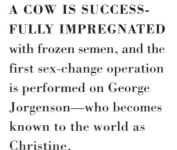

WHILE CROSSING THE ATLANTIC in a salvaged World War II tugboat, a seasick 25-year-old graduate student from Philadelphia named Noam Chomsky first conceives the idea that the biological structure of the human brain genetically predetermines human children to communicate by means of language. Although he has difficulty finding a publisher, four years later his first book, *Syntactic Structures*, completely redirects linguistic research.

INVENTED IN 1948, but initially unreliable, the transistor begins appearing in hearing aids and other devices. Meanwhile, the Japanese are working on transistor radios far smaller and more reliable than any yet developed. The age of miniaturization is on its way.

THE KOREAN WAR ENDS with the formation of a demilitarized zone between north and south.

▶ PRIOR TO A VACCINE, ONE OF EVERY 3,000 PEOPLE IN THE U.S. WAS STRICKENED WITH POLIO.

54

U.S. MICROBIOLOGIST JONAS SALK announces a vaccine against the crippling viral disease poliomyelitis. To create the vaccine, Salk grows all three types of the polio virus to concentrated strength and then kills them. Although impotent, the killed virus successfully stimulates antibody production when injected.

CLINICAL TESTS BEGIN on the first oral contraceptive for women, who will eventually christen it simply "the pill."

55

THE U.S. CIVIL RIGHTS MOVEMENT is galvanized when Rosa Parks, a 42-year old black seamstress, is arrested in Montgomery, Ala., for refusing to relinquish her bus seat to white passengers.

ALBERT EINSTEIN DIES.

56

WHILE MAPPING the earth's ocean basins, scientists discover a belt of mountain ridges and rift fractures that circle the globe much like the seams of a baseball. Their findings will lay the groundwork for the theory of plate tectonics that will develop in the decade to come.

57

THE UNITED STATES IS SHOCKED when the first rocket catapulted into orbit is the Soviet spacecraft Sputnik I. Later that year, a Russian-born dog becomes the first earthling in space aboard Sputnik II.

SEARCHING FOR UNITY

Deep in the hearts of natural scientists lies an urge for unification—the desire to understand a great many apparently different phenomena in simple, unified terms. But the sands of scientific history are white with the bones of failed attempts at unification. Thales of Miletus was wrong when, about 600 B.C., he taught that everything was made of water, and the atomic

Steven Weinberg

taught that everything was made of water, and the atomic theorists of the early nineteenth century were wrong when they supposed that not only the chemical elements, but even heat and light, were composed of atoms.

A great step toward unification was taken by the British physicist James Clerk Maxwell in the 1860s. It was already known that there was a connection of some sort between electricity and magnetism. For instance, a changing magnetic field produced an electric field, which in a dynamo drives electric currents through wires. Maxwell proposed a set of equations in which this relation is reciprocal: according to these equations, also a changing electric field produces a magnetic field. Some of the solutions of Maxwell's equations described a situation where changing magnetic fields produced changing electric fields which in turn produced changing magnetic fields, so that there is a self-sustaining electromagnetic wave traveling through nominally empty space. Maxwell calculated the speed of these waves, and, finding that it was the same as the measured speed of light, concluded that light is nothing but an electromagnetic wave. He had thus succeeded in unifying the phenomena of electricity, magnetism, and light.

◀

UNIFICATION THEORIES SEEK TO EXPLAIN ALL PHENOMENA OF THE UNIVERSE WITH ONE THEORY.

The only other field of force that was known at the time was that of gravitation. After Albert Einstein in 1915 presented his theory of gravitation, the General Theory of Relativity, he turned to the problem of unifying this theory with Maxwell's theory of electromagnetism, in what he called a unified field theory. Unfortunately, this was yet another premature effort at unification, for two reasons. One is that no unification is possible

JAMES CLERK MAXWELL

without taking into account other forces in nature, forces whose existence was not be clearly recognized for decades after the advent of General Relativity. These are the strong nuclear forces, which hold quarks together inside the particles in the nuclei of atoms, and the weak nuclear forces, which cause quarks of one sort to turn into quarks of other sorts, releasing energy that among other things provides the heat of the sun. The other reason why Einstein's unified field theory was premature was that further unification had to be achieved within a dynamical framework that was very different from that of all earlier theories from Newton to Einstein, a framework provided in the 1920s by the discovery of quantum mechanics.

One of the first revelations of quantum mechanics is that the energy and momentum of any field like the electromagnetic field come in tiny bundles, or *quanta*. The quanta of the electromagnetic field are the particles of light first proposed by Einstein in 1905, and later called *photons*. But in the new quantum field theory there was also an electron field, whose quanta are the particles known as *electrons*. Quantum field theory unified force and matter; every sort of elementary particle, it showed, is the quantum of some field, and every sort of force is produced by a field.

The next step toward a unified view of physical force was the unification of electromagnetism with the weak nuclear forces. At first sight, this might seem an unpromising direction for unification. No one had yet observed any particles that might be the quanta of the fields that produce the weak nuclear force, so these particles would have to be very heavy, too heavy, to have been created from the energies available in existing accelerators. (This is part of the reason that the forces produced by these fields are so weak.) In contrast, the photons that are the quanta of the electromagnetic field to have no mass at all. How could such different fields be described in a unified way?

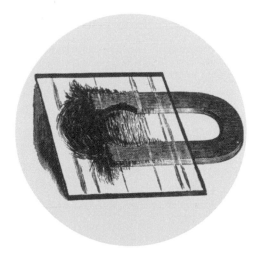

▶

The key to the puzzle was provided by a phenomenon called *broken symmetry*. A symmetry of the laws of nature is a statement that these laws look the same from different points of view. The most familiar symmetries are the symmetries of objects, as, for example, that square rotated 90° looks the same as before it was rotated. What concerns us here is not the symmetries of objects, but of laws, as for instance the symmetry that dictates that physiologists discover the same equations however their laboratories are orientated. It is possible for the equations of a theory to have a symmetry, as, for instance, it may be that these equations may stay the same when we interchange what we call electromagnetism and the weak nuclear forces, while the symmetry is absent in the *solutions* to the equations. We find the masses of particles by solving the equations that govern the fields, and so these masses may not reflect the symmetries of the underlying theory. These ideas led in the late 1960s to a unified theory of weak and electromagnetic forces

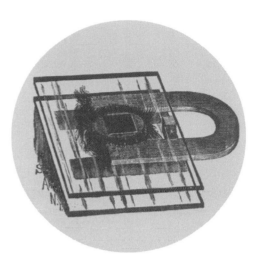

that was not only aesthetically satisfying and mathematically consistent, but that actually worked. The theory predicted new kinds of weak nuclear force that were discovered experimentally in 1973, and new particles that were discovered in 1984.

Having seen the successful unification of the weak and electromagnetic forces, physicists naturally turned their attention to the other two known forces of nature, the strong nuclear force and the gravitational force. Again, this appeared unpromising. For gravitation, it seemed that any quantum field theory would lead to mathematical inconsistencies, that arise from a rapid increase in the force of gravitation at very short distances. For the strong nuclear forces there was a different problem. By 1973, theorists had developed a satisfactory quantum field theory of the strong nuclear force, known as *quantum chromodynamics*. In this theory there are eight additional fields, whose quanta, like the photon, had zero mass. Several attempts were made to formulate theories in which quantum chromodynamics and the unified electroweak theory would emerge from a unified complex of forces. But as its name implies, the strong nuclear force is much stronger than the electromagnetic force, even though both are transmitted by particles of zero mass. It did not seem that this difference in the strength of forces could be explained by any sort of broken symmetry.

There is a possible solution to this problem, suggested by a characteristic feature of quantum chromodynamics, that the strength of the strong nuclear forces *decreases* very slowly with decreasing distance. It may be that at very short distances, the strengths of the weak, electromagnetic, and strong nuclear forces would all be equal. The distances involved are about as small compared with the size of the orbits of electrons in atoms as those atomic orbits are compared with the earth's orbit around the sun. These distances are so short that, although on the scale of an atom or a nucleus gravitation is negligibly weak compared with the

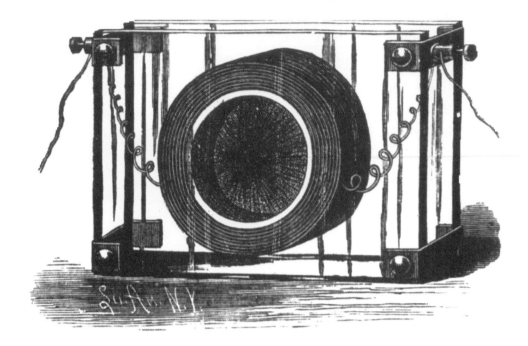

▶

EARLY EXPERIMENTS
ON ELECTRO-MAGNET-
ICS USING SUSPENDED
PARTICLES OF IRON

other forces, at these tiny distances gravitation would be about as strong as any other force. Thus it seems that the future of unification in physics lies at extraordinarily short distances, and that progress is unlikely unless we bring gravitation into the picture along with the strong, weak, and electromagnetic forces.

Today there is just one candidate for such a unified theory of all the forces. It is known as *superstring theory*. According to the general ideas of superstring theory, elementary particles such as electrons and photons and quarks are not point particles at all, but tiny one-dimensional strings in various modes of vibration. Superstring theory is mathematically difficult, and so far has yielded no successful quantitative predictions. On the other hand, not only can superstring theory accommodate gravitation, it makes the existence of gravitation a mathematical necessity. This together with its consistency and beauty, augurs well for future unification. But whether in terms of superstring theory or not, there is no doubt that physicists will continue their historic search for a unified view of nature.

May the Neutral Current Force Be with You

The history of the unified theory of weak and electromagnetic forces provides a nice example of the interaction of theory and experiment. When it was first proposed, the most striking prediction of this theory was the existence of a new kind of weak nuclear force. It is called a neutral current force, because although energy and momentum flows between the particles that exert this force on each other, electric charge does not. In contrast, the familiar weak forces that give rise to radioactivity are called charged current forces because electric charge is transformed among the particles involved. At the time when unified theory was worked out, no neutral current forces had been identified experimentally—from which one could infer either that they did not exist, or were weaker than the ordinary charged current forces. Most physicists assumed that they did not exist. But the new theory predicted that the neutral current forces should be somewhat weaker then the charged current forces, so experimentalists now for the first time set out to find them. In 1973, they reported success, first at the CERN laboratory in Switzerland and then at Fermilab near Chicago. Once this was settled, it was realized that experiments on other issues had in fact been detecting effects of weak neutral currents for years! But those effects were very weak, and without a convincing theory that predicted neutral currents, it had always been assumed that they arose from some sort of contamination in the experiment, and could be safely ignored. These seemingly unimportant artifacts arose from other sources.

THE RESTLESS EARTH

There have only been a few times in the history of science when the introduction of a new way of thinking has transformed an entire field, as genetic DNA did biology, or quantum mechanics did physics. It was the principle of plate tectonics which triggered this revolution in all of the Earth sciences; geology textbooks before and after its articulation bore no resemblance to each other. As recently as the mid-1960s, geologists could not reasonably explain how mountains formed, what caused earthquakes and volcanoes, or why our resources like oil and metal ores exist in some countries and not in others. Plate tectonics, describing the motion of the continents about the surface of the Earth, finally provided the big picture into which would fit the details of Earth's geology.

Michael E. Wysession

Plate tectonics is a great deal more than just drifting continents. The continents, as well as the sea-floor, are only the top level of a stiff layer of rock called the lithosphere. This 80-100 kilometer-deep rigid layer is broken into more than a dozen major pieces, called "plates," which move independently across the globe carrying the continental crust (about 35 kilometers thick) and the oceanic crust (about 10 kilometers thick) along with them. Mountains form where plates collide; with one of the plates usually sliding beneath the other. Where plates move apart volcanic rifts, called mid-ocean ridges, emerge, with molten rock flowing up to fill the resulting gap. On an even bigger scale, the plates, which move at the rate of zero to ten centimeters per year, are only the surface expression of mantle-wide convection patterns. These patterns are caused by the slow, but steady, sinking of cold rock from the surface and the rising of hot rock from the base of the mantle. The mantle is the top half of the Earth, extending from the shallow crust down to the iron core, a depth of 2890 kilometers. Convection occurs when solid mantle rock behaves like a fluid over very long time scales; this activity, not unlike that seen in a boiling pot of water, is an important part of Earth's process of cooling off and losing its internal heat to space.

The idea of plate tectonics had its early roots in the theory of Continental Drift, which was first introduced by the German meteorologist Alfred Wegener in 1915. Wegener noticed that the continents fit together like jig-saw puzzle pieces, and he proposed that the continents had recently been connected as a giant supercontinent, which he called Pangea. Vital to Wegener's theory were the many

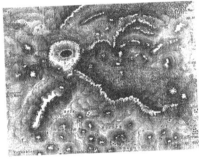

◄ ▲

ON JULY 10, 1892, AN EARTHQUAKE FOLLOWED BY AN EXPLOSION OF MT. ETNA CAUSED CONSIDERABLE DAMAGE TO THE TOWN OF NICOLOSI, SHOWN IN THE MAP ABOVE.

similar geological features, such as land animal fossils, glacial deposits and ancient deserts, which existed simultaneously across continents that were now thousands of kilometers apart. During Wegener's time, Continental Drift was rejected by geologists because no force could be found that could move the continents such great distances through solid rock.

In 1967, the idea of Continental Drift gained a new respectability with the formal establishment of "plate tectonics," determined independently by J. Morgan (US) and the team of Dan McKenzie and Robert Parker (UK). The revival was generated by extensive mapping of the sea floor that had been started during the Second World War and that had been continued through the 1950s, work that had been especially performed by geologists like Harry Hess, who had been a U.S. Navy ship captain. In 1963, Frederick Vine and Drummond Matthews (UK), using magnetic imprints within the oceanic crust, observed that long undersea mountain chains that extend throughout the oceans are spreading out from their centers where new oceanic crust and lithosphere is formed.

An important consequence of plate tectonics is the discovery that continental and oceanic crust are distinct formations, with differing ages, histories and compositions. Ocean crust is created at mid-ocean ridges, moves away from the ridge at rates of one to ten centimeters per year, and is destroyed when it subducts into the mantle at oceanic trenches. Such trenches form the collision boundaries between two oceanic plates, or an oceanic and a continental plate. The oceanic crust is made of a rock called basalt, which flows out of undersea volcanoes at the ridges; because of subduction, the oceanic crust is nowhere more than 200 million years old.

Continental crust, on the other hand, is made of more buoyant rocks, like granite, which have separated from the mantle over long periods of volcanism and never sink back into the Earth. As a result, the interiors of continents are more than four billion years old, and some of the oldest of these rocks are themselves of sedimentary origin, which indicates that they were formed from eroded sediments of even older rocks. Continents are like giant rafts of rock which bounce about the surface and grow in size over time, with new material added either by volcanism or, at the edges, by collisions. The heart of North America, called the Canadian craton, is more than 3.5 billion years old, while the rock on the east and west coasts is only a few hundred million years old—still very young in geological terms.

The boundaries of plates are where the most exciting geological events occur. Recurrent earthquakes, that plague Californians living near the San Andreas fault system, occur because the Pacific plate is sliding past the North American plate in a northwesterly direction at about 5.5 centimeters per year, or at about the rate that your hair grows. The two plates do not slide smoothly past each other. Segments of the transform fault (the San Andreas) that separate them remain locked together until a critical stress level is reached. Then, in a matter of seconds, many miles of plates will suddenly slip past each other, causing an earthquake. The San Andreas is just one of many transform faults that separate plates around the world.

The largest earthquakes have a different source. They occur beneath trenches in subduction zones, where regions of the under-riding plate that can be up to 500 kilometers in length can instantly slide more than 20 meters into the Earth.

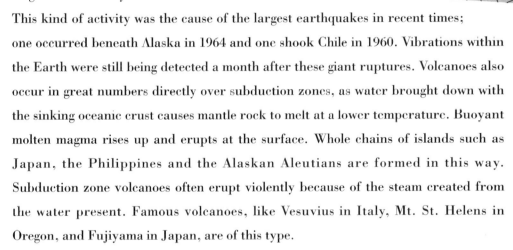

▲

DIFFERENCES IN
LAYERS IN THE CRUST
OFTEN ARE GOOD
INDICATORS OF WHERE
FAULTS CAN BE FOUND.

This kind of activity was the cause of the largest earthquakes in recent times; one occurred beneath Alaska in 1964 and one shook Chile in 1960. Vibrations within the Earth were still being detected a month after these giant ruptures. Volcanoes also occur in great numbers directly over subduction zones, as water brought down with the sinking oceanic crust causes mantle rock to melt at a lower temperature. Buoyant molten magma rises up and erupts at the surface. Whole chains of islands such as Japan, the Philippines and the Alaskan Aleutians are formed in this way. Subduction zone volcanoes often erupt violently because of the steam created from the water present. Famous volcanoes, like Vesuvius in Italy, Mt. St. Helens in Oregon, and Fujiyama in Japan, are of this type.

We also find the greatest concentrations of valuable ores and precious metals at the collisional sites of subductions, as hot water transports and concentrates rare elements. California gold-rush prospectors of the 1800s were digging out the hydrothermal ore deposits that formed when part of the Pacific plate sank beneath California and formed the state's Coastal Range mountains. Large oil and natural gas deposits also form when sedimentary basins are folded and thrust deep into the Earth and are caught between colliding continents. The greatest example of this phenomenon is the Persian Gulf Basin, where one single sedimentary basin holds more than half of the total oil ever discovered in the world. Beginning more than 150 million years ago, Arabia and Iran, which were not yet attached, started to move together and to compress the sedimentary basin in between them. The basin, rich in organic matter, was pushed down into the Earth by one of the collisions that followed the break-up of Pangea and reached the ideal temperatures and pressure levels favorable for the formation of oil. The arched and deformed layers of the basin rock also provided containment for the hydrocarbons and held them in place until they could be drilled and extracted.

Collisions between continents can also be very dramatic, causing giant mountain ranges to be thrust upward. The Appalachian mountains were formed when North America crashed into Africa during the formation of Pangea 300-250 million years ago. The tall Atlas mountains in Africa were formed at the same time, although erosion has removed many kilometers of rock from the summits of both mountain ranges. The most spectacular example of such a collision occurred between India and Asia and created the enormous Himalayan Plateau, which contains all of the world's tallest mountains. The collision began about 50 million years ago, and is still continuing, causing many large earthquakes as rock is slowly squeezed both up and out to the sides to the east and west.

Mid-ocean ridges are the sites of many earthquakes as well as frequent volcanism. Robotic submersibles have shown that these undersea ridge crests, covered with fresh lava, are the sites of active and unusual life forms, including never-before-seen crabs and worms which swarm around smoking chimneys of hot water pumped out of the young crust. Significantly, the birth site of oceanic plates may also be the location where life first began on Earth four billion years ago, as is believed by many biologists and chemists.

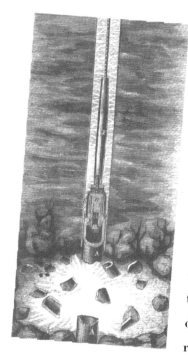

Geologists have back-tracked more than a billion years to follow the paths the plates have taken over time. The necessary information is provided by magnetic sea floor stripes, as well as by continental paleomagnetism and hot-spot locations. It turns out that the direction of the magnetic field reverses randomly, with the time between reversals ranging from 10,000 to 35 million years. This phenomenon causes the magnetic field of the oceanic crust, which is frozen into the rock when it forms at the ridge, to have a striped appearance. The ages of all parts of the ocean crust can be dated by correlating the patterns of the magnetic stripes with a known record of magnetic-field reversals. When volcanic lava hardens and cools on land, it also freezes in the direction and distance to the north and south magnetic poles. Even if the continent rotates and moves relative to the poles, igneous rock retains the ancient magnetic field direction, or paleomagnetism, from the time it first formed. This configuration and the sea-floor spreading record tell us how plates have moved relative to each other. To get an absolute measure of plate motions, relative to the interior of the Earth, hot spot tracks are used.

A hot spot is a rising plume of hot rock, more than 300 degrees Centigrade hotter than the surrounding rock, that rises from deep within the mantle and melts at the surface, erupting as giant volcanoes like Hawaii and Iceland. There are more than 40 major hot spots, which most likely originate at the boundary between the mantle and the core. To a high degree, they are fixed relative to the mantle and to each other. When plates move over them, hot spots punch through the plates, making long chains of volcanic islands that are youngest over the hot spot and get older as they move away. These volcanic chains show the absolute direction the plates have moved. Hawaii is actually the youngest in a long chain of islands and underwater seamounts which extends across the Pacific ocean floor all the way to the Pacific trench under Kamchatka, Siberia. There is even a bend in the volcanic chain near Midway Island, a 43-million year old volcano, which tells us that the Pacific plate changed the direction of its motion 43 million years ago.

The history of the plate motions indicates that not only were all the continents joined together when they formed the supercontinent Pangea, which began to break up 225 million years ago, but that they had previously been joined in another supercontinent 1.2 billion years ago, with a different configuration. In fact, at that time, if any animals had existed (which none did), they could have walked directly from what is now Pennsylvania to what is now Chile in South America. There seems to be a cycle, called the Wilson cycle after geophysicist, Canadian J. Tuzo Wilson, of the continents regularly coming together and then breaking apart.

The history of the movements and collisions of the continents now reveals not only the locations of mountains, earthquakes and mineral resources, but the whole geologic record contained within sedimentary rocks. Plate tectonics is, however, still a very young science, and there are many questions that still remain. We know, for example, that the oceanic lithosphere sinks into the mantle, but there is a debate as to whether the plates stop 650 kilometers down and assimilate into the upper mantle, or whether they continue all the way down to the core-mantle boundary. The prevailing idea is that sometimes they do and sometimes they don't; the matter is far from settled. We know that plate tectonics is related to mantle convection, but we don't know the relative degrees to which the convection moves the plates, or to which the subduction drives the convection. Today, geologists and geophysicists are studying the core-mantle boundary, which is very variable and seems to control much of the dynamics of the interior of the planet.

GROWING BEYOND
OUR LIMITS

In the past 65 million years, perhaps the most stunning event on our planet has been the explosive growth of the human enterprise. From a few hundred or a few thousand upright, small-brained individuals who lived some 3.5 million years ago, the first human creatures evolved into a hugely successful species that

**Anne H. Ehrlich
and
Paul R. Ehrlich**

spread around the world and that increased its numbers to 5-10 million by the time, 10,000 years ago, when agriculture was first invented. The advent of agriculture resulted in increased birth rates and, eventually, to lower death rates. When birth rates are higher than death rates, a population grows. About 150 years ago, slightly over one billion people inhabited Earth.

As the scientific age dawned, improved sanitation methods led to further declines in death rates, followed by slowly falling birth rates in industrializing nations. The two-billion mark was passed around 1930. After the Second World War, the rapid and widespread dissemination of modern health technologies, especially antibiotics and DDT and other pesticides that were used against malarial mosquitoes, caused death rates to plummet in less developed nations. With no compensating attempt made to lower birth rates, this circumstance produced a record spurt of population growth. By 1994, the world population had reached 5.6 billion and is still increasing by roughly 90 million people a year.

But the numbers themselves tell only a small part of the story. Much more important is the impact of human activities on Earth's life-support systems on which our civilization depends. To determine the impact, we must know not only the number of people on the Earth, but also the environmental consequences on the average individual. Everything else being equal, six billion vegetarians who walked or bicycled to their daily activities would destroy their environment much more slowly than six billion steak-eaters who drove gas-guzzling cars several hours a day. In the latter case, the scale of the human enterprise would be far greater than in the former.

That scale can be viewed as the product of three factors—population size, per-capita consumption or affluence, and a measure of the degree to which the technologies used to supply each unit of consumption damage the environment. Since governments don't keep statistics on overall consumption or environmental damage done by a nation's technological apparatus, analysts, such as John Holdren of the University of California, Berkeley, have employed per-capita energy consumption as a useful, although imperfect, substitute for our determining the effects of per capita consumption on technology. Since *Scientific American* was founded, the world's population size has multiplied roughly five times and per capita energy consumption roughly four times; so the scale of the human enterprise and its impact on the environment has increased 20-fold. In the past 150 years or so, population growth has contributed slightly more to environmental disruption than has "affluence" (or more accurately, the damage done by the technologies supplying each person with goods and services).

The almost six billion people living today cannot be supported on the "income" produced by nature; they can only be supported by destroying or dispersing the priceless and irreplaceable stock of "capital" that humanity has been given. The most critical elements of that capital are not, as one might suppose, fossil fuels or other minerals. For the medium term, there is an abundant supply of them—although other environmental considerations, the emission of greenhouse gases from the burning of fossil fuels, for example—are beginning to force us to limit their use.

One critical element of our capital is rich agricultural soils. These are generated by natural systems on a timescale of inches per millennium, and they are being destroyed in many areas at inches per decade. Another threatened resource is ice-age groundwater, which almost everywhere is being pumped from aquifers at many times the rate of recharge. A third is biodiversity: the populations of nonhuman species of organisms, which are integral working parts of ecosystems. The services that ecosystems perform include: maintaining the quality

Overpopulation in the U.S.

Contrary to common wisdom, from the standpoint of human impact on Earth's life-support systems, the United States is the most overpopulated nation. With 260 million people it is the third most populous nation on the planet. There are today 125 million more people than should be living in the United States at the same time according to historic justifications.

Americans are superconsumers, and use rather inefficient technologies to service their consumption. The average American uses almost 12 kilowatts of energy, some 70 percent more than the average citizen of any other industrialized nation. America's total environmental impact is roughly six times that of the 900 million people of India.

of the atmosphere, recycling nutrients, providing a steady flow of fresh water, generating and fertilizing soils, controlling the vast majority of potential crop pests, and maintaining a "genetic bank" from which we have already withdrawn the basic currency of civilization, including crops and domestic animals.

Humanity has clearly expanded beyond the "carrying capacity" of Earth for human beings. That capacity is defined as the number of people that can be supported indefinitely in their present lifestyle. Obviously, even the 5.6 billion people currently living on Earth cannot be supported indefinitely if capital must be exhausted to maintain them today. And the situation is only going to worsen; projections indicate that our population will reach at least 9-12 billion before its growth ends. One international research group suggested that economic activity would need to be multiplied five- or ten-fold in order to provide all members of a doubled population with a decent living. On the other hand, ecologists believe that such a huge increase in the scale of economic activity would wreck the planet's ecosystems, and that any attempt to feed such a multitude is likely to fail completely or at best succeed only temporarily and unsustainably.

Today, the food situation is already marginal. On a per-capita basis, grain production has failed to match the level of 1984, although in absolute terms the trend is still moving upward. Cereals are the foundation of the world's food supply: in rich societies they are used mostly as animal foods rather than as direct nourishment for people. The slippage in grain production of the last decade has been compensated for by a reduction in the demand for meat, and so has gone largely unnoticed. But many observers are concerned that the green revolution, which kept food production ahead of population growth between 1950 and 1985, has run its course, and there are no technological miracles in sight to replace it.

Holdren has outlined the most optimistic solution to the human dilemma. Over the next century, with a sufficient effort, population growth could be halted at ten billion; and by becoming more efficient, rich nations could reduce their per-capita energy use from 7.4 kilowatts (kW) (a continuous flow of energy equivalent to 74 100-watt lightbulbs) to 3 kW. At the same time, poor nations could increase their energy use per person from 1 kW to 3 kW, effectively closing the rich-poor gap. These actions would give us a world with roughly 2.3 times its present level of economic activity, as measured by energy use. Whether such a scale of activity could be supported even temporarily until the human enterprise could be shrunk to a level maintainable in perpetuity is not now clear. At the very least, it would require a level of care in the treatment of environmental systems that we have not yet approached.

THE SPACE AGE

▶

1960

THE UNITED STATES LAUNCHES the first weather satellite, Tiros I, greatly improving meteorologists' ability to predict the weather.

MARY AND LOUIS LEAKEY
IN TANZANIA
▼

61

SOVIET COSMONAUT YURY GAGARIN becomes the first human in space on April 12. His spacecraft, Vostok I, orbits the Earth once during its 108-minute voyage. U.S. astronaut Alan Shepard, Jr., follows Gagarin on May 5 with a 15-minute suborbital flight in his capsule Freedom 7.

HOMO HABILIS, THE FIRST HOMINID capable of shaping stone tools, is discovered by British anthropologists Louis and Mary Leakey in Olduvai Gorge, Tanzania. The species, an evolutionary step between Australopithecines and Homo erectus, walked the earth some 1.8 million years ago.

62

IN HER BOOK SILENT SPRING, U.S. biologist Rachel Carson delivers an impassioned account of the potentially devastating effects of pesticides on the planet's wildlife.

THE FIRST COMMUNICATIONS SATELLITE, Telstar I, is launched. The global village is born.

U.S. PHILOSOPHER AND HISTORIAN of science Thomas Kuhn publishes his influential book *The Structure of Scientific Revolutions,* which likens the struggle to change scientific paradigms to a political revolution in which the bloodshed is intellectual rather than physical.

AS THE VIETNAM WAR INTENSIFIES, the United States sends its first military advisers to assist South Vietnamese forces. Direct U.S. involvement in the Vietnam War begins.

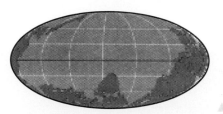

600 MILLION YEARS AGO.

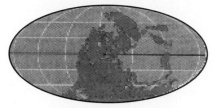

300 MILLION YEARS AGO.

100 MILLION YEARS AGO.

63

QUESTIONS THAT SEEM UNANSWERABLE under the old paradigm, the development of more sophisticated geophysical techniques and the amassing of new data lead to a revolution in earth science—the formulation of plate tectonic theory. The earth's crust is divided into plates that are driven about the planet's surface by the convection of molten material in the mantle below. The interactions where plates meet account for most of the earth's dynamic processes such as volcanism, earthquakes, and mountain-building.

65

USING AN ELECTRON MICROSCOPE to analyze tiny flecks of carbonized material in ancient rocks, scientists discover the fossilized remnants of simple cells. Found in rocks as old as 3.5 billion years, these microfossils indicate that life on earth began far earlier than was previously believed—as few as one billion years after the planet's birth.

TWO U.S. PHYSICISTS, Arno Penzias and Robert Wilson, worry when their ultrasensitive microwave detector senses background noise. Present day and night, season after season, they conclude that the microwaves are emanating from beyond the Milky Way galaxy. They have discovered the energy left over from the big bang that formed the universe.

U.S. BIOCHEMIST Robert Merrifield successfully synthesizes the protein insulin.

67

THE FIRST SUCCESSFUL HEART TRANSPLANT is performed by South African surgeon Christiaan Neethling Barnard.

69

U.S. ASTRONAUT NEIL ARMSTRONG takes his "giant leap for mankind" when he descends the steps of the Eagle lunar module and becomes the first human to walk on the moon. Armstrong and his partner, Buzz Aldrin, spend two hours on the lunar surface gathering rock samples, setting up experiments and romping in the weak lunar gravity. Their television camera beams the event a quarter of a million miles to an amazed earth audience.

BUZZ ALDRIN WALKING ON THE MOON
▼

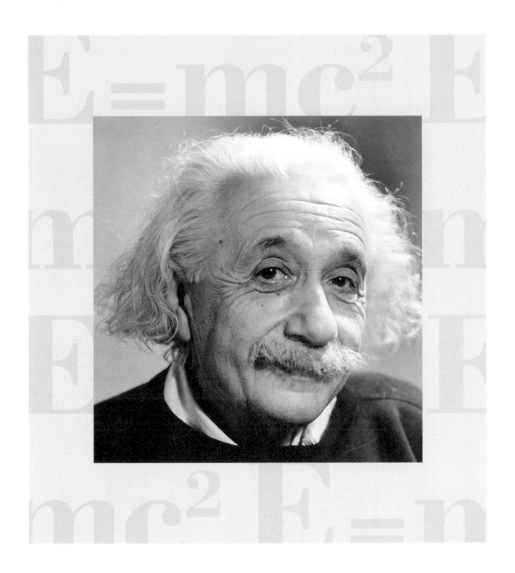

THE THEORY OF GRAVITY

Isaac Newton was the most admired scientist of his time. When he produced his semi-nal treatise, the *Mathematical Principles of Natural Philosophy*, he was rewarded by being appointed Master of the Royal Mint, but he had already transformed

Joseph Silk

gravity from a mythological fancy into a reality of nature. Action-at-a-distance could account for the moon's passage around the earth, and for Johannes Kepler's laws that described the orbital motions of the planets around the sun. But there was still an element of mystery and intangibility in the story, for what medium transmitted the force of gravity?

The American physicist Albert Michelson and his chemist colleague Edward Morley searched in vain for the elusive ether, the hallmark of Newtonian absolute space. There was no indication that the speed of light varied with the direction of motion of the earth. This result rocked the very underpinnings of Newtonian absolute space. Most scientists regarded any tinkering with Newton's gravity as heresy, and preferred to wait until further experiments confirmed the status quo. It required the genius of Albert Einstein to understand that the Michelson-Morley experiment heralded the overthrow of absolute space. Einstein reconciled experiment and theory by relating gravity and space in a fundamentally new way.

Geometry provided the key. Gravity equals geometry, said Einstein. Where there is gravity, space is distorted. Wherever space is curved and Euclid's axioms fail, there is gravity. No longer was there any confusion about how gravity operated: everything to do with gravity was in the geometry of space. Even on earth, space is curved. The geometry taught at school is flawed. Not by much, only one part in a billion, but it is the principle that is teaching us something new and fundamental. And there are regions of space, so Einstein's theory predicts, where the curvature is far more extreme. Near a black hole, the geometry is so distorted that no material object would be strong enough to resist the stresses and strains of its warped space. These are tidal forces, innocuous enough in the earth's oceans where tides are our most direct encounter with gravity, but irresistible around a black hole.

The prediction that black holes should exist was implicit in Einstein's theory, although acknowledgment among scientists of their likely existence was not forth-coming for several decades. They didn't even have a name until "black hole" was coined in 1967 by the American physicist John Archibald Wheeler. This was partly in response to the discovery of quasars, immensely luminous, compact and remote objects, which were conjectured to be powered by supermassive objects in the nuclei of galaxies. In fact, even Newton came close to realizing that such objects should exist. The prediction of objects where the gravity field was so great that light could not escape was first made in 1783 by John Michell and independently by Pierre-Simon Laplace in 1795.

While Newton's theory may allow black holes to exist, Einstein's theory led to profoundly new insights. Black-hole formation is generally associated with gravitational radiation. Gravity-wave experiments are underway in at least three continents, and are under study via the format of a network of space satellites. Black holes contain singularities, which are generally inaccessible, shrouded by an event horizon. But black holes

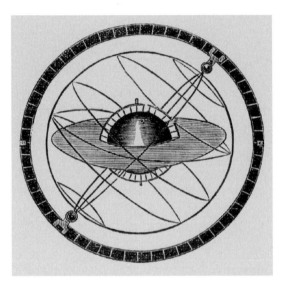

eventually evaporate, perhaps exposing singularities where the laws of physics may break down, to unleash the secrets of quantum gravity, the long-sought unification of gravitation and quantum theory that physicists are eagerly pursuing. These prediction are still far from being fulfilled. However, relativity theory has successfully confronted experiment via at least three distinct challenges.

Pride of place goes to orbital precession. Explaining the precession of Mercury's orbit around the sun was regarded by Einstein as a major achievement of his theory of gravitation. However this apparent success was disputed when American physicist Robert Dicke insisted that the solar interior might be slightly oblate, thereby providing the necessary torque to precess Mercury's orbital axis. The discovery of the pulsar pair PSR 1913+16, a binary system consisting of two orbiting neutron stars, provided fertile ground for a more precise test of Einstein's theory. The pulsars, magnetized, rapidly spinning neutron stars, emit highly regular periodic radio beams that provide very precise and stable orbiting clocks. Indeed, the accuracy of a pulsar clock compares favorably with the best terrestrial time standards. The motion of the neutron stars can be measured as the pulsar radiation is Doppler-shifted over a series of many orbits.

Orbital precession has been measured in the binary neutron-star system that amounts to four degrees per year. This is incontestably an effect of general relativity.

Moreover, the orbits are found to be decaying because of the emission of gravitational radiation, also as predicted by Einstein's theory. The Nobel Prize was given in 1993 to Joseph Taylor and Russel Hulse for their description of the binary pulsar as a laboratory for studying gravitation theory.

A second prediction of Einstein's theory was that the paths traversed by light rays would bend on passing through a gravity field. Two expeditions set off to South America soon after the end of the First World War to monitor stars near the sun during a total eclipse. The prediction was that the geometry of curved space near the sun would cause the apparent positions of stars to be displaced by twice the value that Newton's theory predicted. Front-page newspaper headlines around the world announced the resulting confirmation of general relativity.

Modern astronomy has taken the bending of light rays a giant step forward. The gravity field of remote galaxies acts as a gravitational lens that bends the light from background quasars and galaxies. Viewed symmetrically, the light from a star-like object seen through a simple gravitational lens would spread out like a ring, the Einstein ring. There are complications that arise if the lens is extended, as would be the case for a

cluster of galaxies that can produce multiple images. If the background galaxy is viewed through the lens from an off-center direction, instead of a ring, the amplified background image is arc-like. All of these phenomena have been discovered in the past decade: multiple quasar images that are split by the simple lens of a massive galaxy, and arcs and arclets that are produced by the extended lens of an intervening cluster of galaxies. Gravitational lensing is now being used to map out the dark-matter distribution in galaxy clusters.

Yet another consequence of light-bending has been a phenomenon dubbed gravitational microlensing. Dark matter in our galaxy halo, if in the form of compact objects of stellar mass (MACHOs), lenses the light from distant stars. The image separation is only a one-thousandth of a second and is unmeasurably small. However, the light is amplified as the orbiting halo MACHO passes in front of the background star. The chance of such an intervening object producing the amplification is infinitesimal, only one part in a million. Hence millions of background stars must be studied in order to find such microlensing events. The amplification is symmetric in time, achromatic, and of course completely independent of the type of star. The duration provides a measure of the MACHO mass. Experiments are underway that are monitoring some ten million stars in the Large Magellanic Cloud, as well as millions of stars in the central bulge of our galaxy. Microlensing events have been discovered that point to MACHOs of mass around 0.1M; whether these are sufficiently numerous to account for the dark matter is not yet known.

Einstein also predicted that light emitted within a gravity field could be observed from afar to be redshifted in wavelength, photons losing energy as they climb out of the gravity field. This was a prediction of the principle of equivalence,

▶

ACCORDING TO
THE PRINCIPLE OF
EQUIVALENCE OF
GENERAL RELATIVITY
THEORY, A CLOCK
WILL SLOW DOWN
AS A CONSEQUENCE
OF THE CURVATURE
OF SPACE.

fundamental to general relativity, according to which a clock slows down as a consequence of the curvature of space. The gravitational redshift has been measured for the earth, where the difference in frequency between light emitted at the top of a laboratory shaft was more by about one part in a billion than light emitted at the bottom. The effect has also been measured on the sun, where it amounts to one part in six million, and for white dwarf stars that are one-hundredth the radius of the sun, where it is one part in ten thousand.

Perhaps the most spectacular application of the gravitational redshift was predicted by Rainer Sachs and Arthur Wolfe in 1967, and measured by the COBE satellite in 1992. Density fluctuations, laid down a mere 10^{-35} second after the big bang, are believed to seed the large-scale structure of the universe. The cosmic microwave background radiation was last scattered by matter when the universe was about three hundred thousand years old, and has subsequently traveled freely to be observed as the primordial fireball radiation at a temperature of 2.74 Kelvin. Its almost perfect blackbody spectrum was generated early in the dense, hot phase of the universe. The density fluctuations were present when the radiation was last scattered, and produce infinitesimal gravitational red (and blue) shifts of the microwave background photons, corresponding to upwards (and downwards) deviations from the mean density. These temperature fluctuations were measured, at a level of one part in one hundred thousand, at almost precisely (to within a factor of two) the level inferred from the present-day cosmic structure, if the structure grew from gravitational instability acting as primordial density fluctuations as the universe expanded.

Einstein's theory of general relativity has now been vindicated in many ways, leaving any rival theories to debate the finest of fine print. In terms of the so-called Newtonian parameters that describe any differences between Einstein's theory and observations, certification is now at a level approaching 95 percent. The binary pulsar has so far provided the primary laboratory for such tests. As time passes, one may expect that continued pulsar monitoring will refine the match of theory and observation to ever-increasing levels of precision.

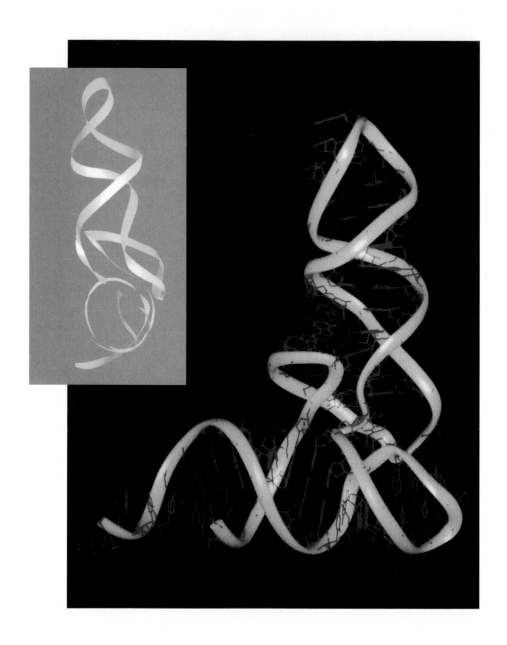

GETTING THE MESSAGE

Cellular instructions are stored in the double helix of DNA. By what means are these instructions transmitted and decoded? The human immunodeficiency virus (HIV) is the causative agent of AIDS, our number one public-health crisis. What is the molecular makeup of the infectious component of this virus? Four billion years ago, the first self-reproducing system arose on the primitive Earth, a harbinger of the origin of life. What molecule is able at the same time to carry information and to perform chemical transformations, and is therefore the prime candidate for such

Thomas R. Cech

a self-reproducing molecule? The answer to all of these questions is the same—RNA.

As one might guess from its three initials, RNA (ribonucleic acid) is chemically similar to DNA (deoxyribonucleic acid), the stuff of genes. Both molecules are polymers, strings of individual building blocks called nucleotides. Each nucleotide unit of RNA contains an extra oxygen atom relative to a nucleotide of DNA; hence the "deoxy" in DNA's chemical name. The extra oxygen atoms in RNA affect its structure and enhance its reactivity. There is another major difference between the two nucleic acids: DNA is found in cells as a double-stranded helix, whereas RNA is produced as single strands. Unlocked from the rigid railroad track of a double helix, RNA strands are free to assume unique, globular shapes that perform any of a myriad of roles in living cells.

◄

THREE-DIMENSIONAL STRUCTURE OF TRANSFER RNA

Most of the fundamental structures and biocatalysts of living cells are proteins. For example, myosin is a key protein in muscle contraction, and pepsin is a stomach enzyme that speeds the breakdown of food. The instructions for making each protein are stored in a chromosome in the form of DNA. The information is transcribed, bit by bit, from the DNA into RNA, which acts as a messenger to carry the information to the ribosome, the cell's protein factory. In addition to messenger RNA, two other classes of RNA, transfer RNA and ribosomal RNA, participate in protein synthesis. The order of building blocks along the DNA specifies the order in which amino acids are strung together to make the protein, which in turn determines the protein's function.

DNA is like an archival master copy of a videotape, which can be copied over and over again. The RNA, like an inexpensive copy of a videotape, may be more short-lived but nevertheless has the same informational content. In order for the information to be accessed, a reading head is required: the VCR machine or, in a living cell, the protein synthesis factory (ribosome). Finally, one can view the image, which in cellular terms is the protein.

Although there were hints in the 1940s that RNA was the genetic intermediary, the role of messenger RNA was not established until 1961. Even then, the genetic code that told which sequence of RNA building blocks specified which amino acid had not been cracked. By 1966, the British scientist Francis Crick, whose ideas inspired much of the thinking about the coding problem, was able to summarize the whole story in his landmark *Scientific American* article, "The Genetic Code: III."

The structure of transfer RNA, which holds an amino acid at one end and recognizes the corresponding code in the messenger RNA at its other end, was soon visualized in atomic detail by Alex Rich and Sung-hou Kim of the U.S.

If DNA makes RNA which makes protein, then you should be able to travel along any of these molecules and read the same message.

WHICH CAME FIRST, THE GENETIC MATERIAL . . .

Returning to the videotape analogy, the first meter of the tape should specify the first minute of the final theatrical production, the second meter the second minute, and so on. That would always be the case. Or would it? In 1977 it was discovered that, especially in humans and other organisms with nucleated cells, many genes are "split"—their information is interrupted by stretches of noncoding DNA dubbed introns. The script of the theatrical production is interrupted by lengthy commercials,

---

A Pinch of Salt

"We isolated unspliced RNA from nuclei of the the single-celled pond organism Tetrahymena. Much to our surprise, simply mixing this RNA with salts and other small molecules found in all cells resulted in the cutting-joining reaction diagnostic of splicing. Such a chemically difficult reaction between very unreactive molecules certainly had to be catalyzed. But what was the catalyst? Our first hypothesis was that the splicing activity was caused by a protein tightly bound to the RNA we had isolated. It would have to be a very unusual protein-RNA complex to survive the multiple forms of abuse to which we had subjected it during RNA purification, which included boiling in the presence of detergent. That we took this hypothesis seriously shows how deeply we were mired in the prevailing wisdom that said that only proteins were capable of highly efficient and specific biological catalysis. It took another year for us to adopt and ultimately prove the alternative, that the RNA chain could fold to form the active site for its own splicing. We coined the term "ribozyme," for a ribonucleic acid with enzyme-like properties."*

Adapted from the Nobel Lecture by Thomas R. Cech, *Les Prix Nobel 1989*, © The Nobel Foundation, 1990

but these are conveniently edited out in the final production. In the cell, this editing takes place at the RNA level. The RNA initially copied from DNA is accurately cut and spliced to remove the introns, so the final messenger RNA has a sequence of nucleotides directly corresponding to the sequence of amino acids in the protein it encodes. How is RNA splicing accomplished? [See sidebar]

Nor is RNA catalysis is not limited to RNA splicing. An RNA processing enzyme, Ribonuclease P, was found to have a catalytic subunit composed of RNA. Furthermore, evidence is mounting that the ribosomal RNA inribosomes is the catalytic heart of the protein synthesis apparatus.

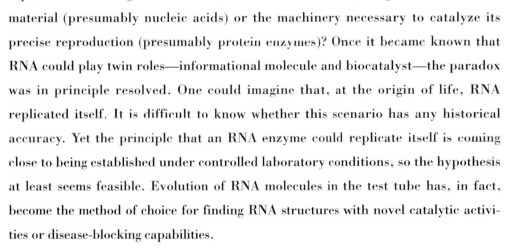

Unknown to many of those working on RNA splicing, there were scientists in another field waiting patiently for the discovery of RNA catalysis. Those researching chemical evolution and the origins of life had come up against a chicken-and-egg problem: if life requires reproduction of the genetic material, then which came first, the genetic material (presumably nucleic acids) or the machinery necessary to catalyze its precise reproduction (presumably protein enzymes)? Once it became known that RNA could play twin roles—informational molecule and biocatalyst—the paradox was in principle resolved. One could imagine that, at the origin of life, RNA replicated itself. It is difficult to know whether this scenario has any historical accuracy. Yet the principle that an RNA enzyme could replicate itself is coming close to being established under controlled laboratory conditions, so the hypothesis at least seems feasible. Evolution of RNA molecules in the test tube has, in fact, become the method of choice for finding RNA structures with novel catalytic activities or disease-blocking capabilities.

. . . OR THE MACHINERY NECESSARY TO CATALYZE ITS PRECISE REPRODUCTION?

It would seem that nothing could be less practical, or less likely to impact on humankind in general, than origin of life research. Yet, the same sort of ribozyme that could mediate its own reproduction could also specifically cleave and thereby inactivate other RNA molecules. Now, consider that many viruses—for example, those that cause the common cold, polio, hepatitis, and AIDS—carry RNA as their genetic information. And all viruses rely on specific messenger RNAs to carry out their infectious cycle. In the test tube, ribozymes have an uncanny ability to seek out and destroy viral RNAs. A prospective future treatment is to deliver these disease-eradicating ribozymes to the relevant human tissues to achieve pharmaceutical efficacy. To make this possibility a reality is one of many challenges facing RNA research in the next decade.

EXTENDING THE MIND

Almost 150 years ago, the English mathematician John Couch Adams and the French mathematician Urbain Leverrier were separately studying the observed but unexplained perturbations of the orbit of Uranus. Each independently

**Piet Hut
and
Gerald Jay Sussman**

hypothesized a transuranian planet and varied the parameters of the hypothetical planet's orbit until a satisfactory reconstruction of the unexplained perturbations was found. Thus Adams and Leverrier predicted the existence and approximate position of a planet beyond Uranus, which was subsequently found near the predicted position and was given the name Neptune.

When Arno Penzias and Robert Wilson discovered the cosmic background radiation at Bell Laboratories in 1965, it was understood that the radiation would not be completely uniform. To seed the eventual for-

mation of structures, such as the galaxies and galaxy clusters that we see today, there must have been some small inhomogeneity in the mass distribution of the early universe that we would see as small fluctuations in the distribution of the radiation. Since Penzias and Wilson, cosmologists have simulated structure formation under a variety of assumptions, and have determined distributions of initial inhomogeneities that could lead to a universe that looks like the one we live in. Recent observations, first with the COBE satellite, have finally detected the initial fluctuations. The sizes and scales measured are now being combined with the simulations to rule out some theories of the composition of the dark matter in the universe.

The discoveries of Neptune, and of the cosmic-background fluctuations, illustrate how computing brings together theoretical and experimental science. Based on previous observations, theorists develop hypothetical models from which predictions can be made that can be tested by observation or experiment. In the past, this synthetic approach was limited to comparatively simple situations. The availability of high-speed computation has allowed synthetic methods to take their place firmly next to the traditional methods of reductionist analysis.

◀

SCIENCE COMPUTATION
CAN IMITATE NATURE.
HERE A LOGARITHMIC
FUNCTION YIELDS A
SMALL SNAIL SHELL.

▶

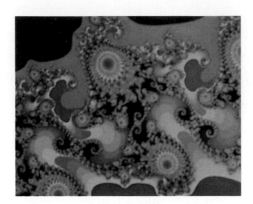

Synthetic computational modeling now plays an important role in diverse areas. Geologists look to computer models to gain insight into tectonic processes and into the origin of the earth's magnetic field. Biologists investigate the consequences of alternative hypotheses of philogenetic trees on the evolutionary history of organisms. Cognitive psychologists now study synthetic computational models in their quest to understand the mechanisms of behavior and language. And economists use synthetic models to understand the consequences of alternative policies.

But besides modeling, there are other ways computation is essential. Computation is needed in the acquisition, processing, and interpretation of data. Scientific visualization, including the new field of virtual reality, is an essential tool now at the command of scientists. And with the rapid growth of scientific knowledge, database and network technology have become crucial for access to, and dissemination of, new results.

A great deal of scientific computation is devoted to signal processing. In experimental and observational data the valuable information is often obscured by "noise." We can use powerful mathematical techniques to extract the signal of interest from the confusion of irrelevant signals and noise. For example, in particle physics, huge numbers of events must be filtered in order to find the few that will give us insight into the phenomena of interest. But even after the information is separated from the noise, the data may not be in a useful form. Data must be transformed to make sense of the information produced by instruments like MRI or ultrasound machines, or from an array of radio telescopes (such as the VLA). To support the massive computation required for data acquisition and scientific visualization, signal-processing systems and graphics engines often contain supercomputers.

Many scientific enterprises, such as the Human Genome Project, depend upon the accumulation and correlation of vast amounts of data collected independently by many investigators. Each contributor supplies sequence fragments, which must be pasted together ultimately to form a coherent picture. The task of reconstruction is a massive combinatorial problem, which has required novel algorithms. Computation to support databases and network communication form a foundation for such cooperative enterprises.

The introduction of new tools can revolutionize a field. Galileo's development of the astronomical telescope changed the way we look at, and think about, the universe. But the diversity of application of computers is even more revolutionary.

One way to think about tools is that they are extensions of the human body. A hammer or a pile driver allows one to exert an impulsive force, greatly in excess of what can be achieved with the unaided arm. An astronomical telescope gives the eye a huge aperture. Each of these is a specialized device, designed to extend a particular sense or effector organ. Paper and pencil are extensions of the mind; in particular, of the memory. A computer is different; it is a general tool. Any computer can be configured, by an appropriate program, to implement any information process that can be described formally. This idea was first formulated independently by Alonzo Church at Princeton and Alan Turing at Cambridge: We say that a computer is a Turing-universal machine. The universality of computers explains the diversity of their current applications, and it ensures that new, qualitatively different, and unexpected applications will continue to appear.

The dominant technology of the nineteenth century was mechanical: the control of power for transportation, construction, and manufacturing. The advent of the twentieth century signaled a transition to the dominance of the processing of information, for the new electricity-based communication technologies, such as telephone, radio, and television. The computer revolution of the second half of the twentieth century is the continuation of that trend. We are now seeing the merger of computing with earlier technologies. We find computing engines embedded in other mechanisms, from microwave ovens to smart measuring instruments and adaptive optics. In the twenty-first century we will see the rise of nanotechnology and its close relative biotechnology, eventually leading to real symbiosis between biological and technological beings.

The Digital Universe

Consider an observation of a galaxy by an array of radio telescopes. First, computers are used to point the telescopes and coordinate the reception of the raw data. Then, a computer transforms the raw data into a recognizable image. The picture is cleaned of noise, and foreground stars are subtracted. And finally, the galaxy is isolated and classified according to its essential qualitative and quantitative properties: an Sb galaxy, at a redshift of 0.4, with a rotational velocity of 240 km/s at a distance of 10 kpc from the center, etc.

THE NEW ENGINEERING

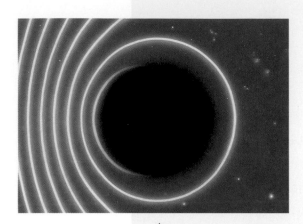

▲

LIGHT RAYS BENT BACK
BY THE INTENSE GRAVITY
NEAR A BLACK HOLE

1970

FIBER OPTIC TECH-NOLOGY is developed to conduct light through fine, flexible clear glass wires. By carrying sound waves as light pulses, the fibers prove capable of transmitting enormous amounts of information.

GENETIC ENGINEERING BEGINS when U.S. microbiologists Hamilton Smith and Daniel Nathans cut a DNA strand with an enzyme and combine one of the fragments with a different DNA fragment to form new genes that do not exist in nature.

71

A STAR IN THE CONSTELLATION OF CYGNUS is detected revolving around a body that radiates X-rays but not visible light. Astronomers conclude the body must be so unimaginably small and dense it generates a gravitational pull not even light can escape. They have discovered the first black hole.

73

MARIO MOLINA, A YOUNG SCIENTIST at the University of California, Irvine, performs calculations to determine the fate of chlorofluorocarbons (CFCs) in the atmosphere. A few months later Molina and his academic advisor U.S. chemist Sherwood Rowland publish chilling evidence that CFCs rise to the earth's upper atmosphere and destroy its fragile ozone shield.

A CEASE-FIRE ENDS the conflict in Vietnam, at least for the United States.

74

ALTHOUGH THE ORIGIN OF THE MOON has puzzled people for millennia, rock samples gathered by the Apollo program cause a resurgence of speculation. Many scientists now believe U.S. astronomer William Hartmann's theory that in the early days of the solar system the moon coalesced from debris generated by a glancing collision between Earth and a Mars-sized planet.

U.S. PALEONTOLOGIST DONALD JOHANSON discovers a three-and-a-half-foot-tall female hominid, Australopithecus afarensis, who walked on the earth three million years ago. Inspired by the popular Beatles' song *Lucy in the Sky with Diamonds*, he names her Lucy.

75

TRANSISTORS are now so small and the circuits on them so compactly etched they are referred to as microchips.

THE FIRST PERSONAL COMPUTER, the Altair 8800, is introduced in kit form in the U.S. Equipped with only 256 bytes of memory, it has no keyboard and must be accessed by flipping switches. Over the next 15 years, however, descendents of the Altair will appear in one of every five American homes.

76

TWO PROBES, VIKING 1 AND 2, land on the Martian surface. They discover that the planet's atmosphere is chiefly carbon dioxide with small amounts of nitrogen and argon. Results of experiments to determine the presence of microscopic life forms are ambiguous, but no trace of organic matter is found in the soil. The dry riverbeds and tributaries indicate there was once liquid water on the planet.

▶ CAST OF DINOSAUR THIGH BONE (MEASURING 6 FEET, 2 INCHES) AS IT APPEARED IN *SCIENTIFIC AMERICAN* IN 1893

78

SCIENTIST ROBERT WEINBERG and his colleagues give tumors to mice simply by transferring individual genes. These newly formed oncogenes, they discover, differ from normal genes in only a single amino acid. Weinberg's team concludes that a chance mutation during a single replication of a normal gene can create a deadly oncogene.

THE FIRST HUMAN TO BE CONCEIVED in a test tube and successfully brought to term—a healthy baby girl—is born to a woman in Great Britain.

79

A NEW THEORY for the extinction of the dinosaurs emerges when U.S. geologist Walter Alvarez and his father, physicist Louis Alvarez, analyze 65-million-year-old Italian soils and find anomalously high levels of the element iridium. Because iridium is rare to the earth but common to meteorites, they theorize that a mass extinction was caused by an asteroid crash that triggered volcanic eruptions, tidal waves and clouds of dust that blocked out sunlight and caused unsurvivable climatic changes.

THE OLDEST OBSESSION

Imagine a world where everything specific about sex is denied, obscured, and covered up, but where sexual excesses flourish. Eminent physicians write sex manuals claiming that sexual restraint promotes better health, while some respectable husbands admit that they think constantly of women and vaginas, and supposedly sexless wives plan affairs and know how to get an abortion if they need one. Parents and experts alike say that children are not sexual beings, yet worry that kids will engage in sex play. A few parents even seduce their own children. Prostitution secretly flourishes, and shameless "gay" women walk the streets.

James D. Weinrich

This world has no talk shows, no international telephone calls, and no nudist camps. This world has never heard of testosterone, the pill, chromosomes, transsexuals, or gay rights ("gay," in fact, is a slang term describing prostitutes). This is the English-speaking world of 150 years ago, into which *Scientific American* was born.

The first 50 years of life of *Scientific American* fell neatly within the Victorian era. Scientific understanding of sexuality relied upon ancient theories of vital humors (fluids), bloodletting, and semen conservation. By the turn of the century, however, a scientific sexual revolution was underway.

Published in 1886 was *Psychopathia Sexualis* by Richard von Krafft-Ebing, a thick book that discussed myriad sexual oddities, such as a woman erotically aroused by putting on men's clothes or a man attracted to women's gloves. In the same era, Sigmund Freud shocked the world by proposing that adult sexual proclivities had their roots in those allegedly sexless childhoods. Even what we know today as the gay-rights movement was born then, as Karl Heinrich Ulrichs campaigned to legalize homosexuality in his native Germany. His view that homosexuality is biologically determined remains controversial today.

Havelock Ellis' book *Studies in the Psychology of Sex*, published in the late 1800's de-pathologized masturbation, orgasm, sexual pleasure between husbands and wives, and homosexuality. Robert Latou Dickinson, an American physician, believed that doctors shouldn't pay attention to the sexual histories, sexual anatomy, and sexual physiology of their patients. Clelia Mosher showed that married women did enjoy orgasms, and that corsets and other restrictive Victorian clothing affected menstruation and distorted the body's internal organs. Magnus

Hirschfeld of Berlin founded the first sex-research journal and the first academic sexological institute (which was later raided and destroyed by the Nazis).

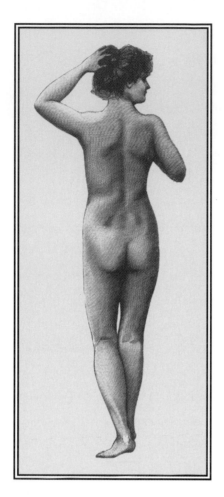

Margaret Sanger fought to make birth control legal, opened the first birth control clinic in the United States, and founded the organization now called Planned Parenthood.

The new century had barely begun when the world's first nudist colony opened in 1903 in Germany. Although nudism per se is not a sexual proclivity, sex and nakedness have always been associated, and some citizens, at least, were now dealing with these issues out in the open. Advances in physics and chemistry were preoccupying much of scientific America during these years, and the study of the chemistry and physiology of sex had its own hormonally induced growth spurt. The first sex hormone, the female hormone estrone, was isolated in the late 1920s; the major male hormone testosterone was discovered in the early 1930s, and progesterone was found in 1934. Testosterone was first synthesized in 1939. Chromosomes (the X-shaped bodies containing DNA in the nuclei of cells) had been discovered in the 1890s, but their association with heredity was not established until 1902. Scientists soon demonstrated that X and Y chromosomes determined the sex of a baby.

But most American scientific researchers went to war during these years—either to World War I or II, and sex-related studies were shelved or abandoned. Even sex research with military applications, such as the use of to fight syphilis and gonorrhea, was difficult to conduct.

The modern post-war era of sex research began with a bang with the publication of Alfred Kinsey's massive studies of sexual behavior in the human male (1948) and female (1953). No one before or since has interviewed so many people so thoroughly about their sex lives; about 17,000 men and women ultimately took part in the studies. Kinsey shocked America when he wrote that 50 percent of American men had been unfaithful to their wives, with 25 percent of

American women returning the favor. He also showed that homosexual experience was far more common among Americans than had been imagined. Although Kinsey's work is often cited purely for its volume of statistics, it is the patterns he found within those numbers that are his lasting intellectual contribution to the field.

Coincidentally, 1953 also saw the publication of the first issue of *Playboy* magazine, with Marilyn Monroe as its centerfold; the first issue of *Playgirl* appeared in the mid-1970s.

Hormone research continued briskly after World War II. Among other discoveries, we learned that, even before birth, sex hormones "organize" circuits in the brain and permanently affect adult sexual life.

Until the 1960s, no one knew how vaginal lubrication happened, and there had been little study of the physiology of sexual arousal and orgasm. So if a woman told her doctor that intercourse was painful, or if a man suffered a splitting headache at the moment of orgasm, therapy was prescribed in ignorance. It fell to William Masters and Virginia Johnson, whose work first appeared in 1966, to discover the source of vaginal lubrication, measure the blood-pressure increases resulting from sexual excitement, establish a formal system of sex therapy, and make the study of such phenomena medically respectable.

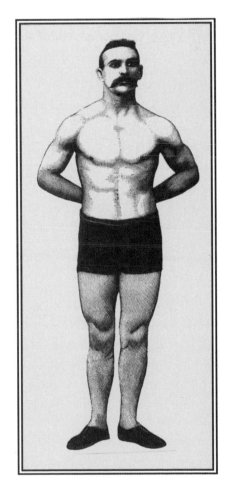

In the 1960s, demystification began illuminating even darker corners. Transsexualism was recognized as something quite different from homosexuality, and transsexual surgery (first performed in the 1950s) improved. Scientists such as John Money and Robert Stoller studied paraphilias: a class of sexual turn-ons for which one's specific partner is less important than a "scene" or fantasy that is acted out. Just like any sexual activity, paraphilias range from the harmless, such as shoe fetishism, to the horrifying, such as lust murder. The physical abuse of children probably has a sadistic-erotic association as well.

As the huge "baby boom" generation became sexually active, the sexual revolution of the 1960s and 1970s began. For the first time in history many college

Masturbatory Insanity

Of all the bizarre beliefs that learned men and women have held about sex, surely the notion of masturbatory insanity is one of the oddest. A century before Scientific American was born, a Swiss physician and popular health writer named Samuel Tissot wrote a book entitled Onanism: Or a treatise upon the disorders produced by masturbation. Tissot wrote that masturbation led to insanity because it wasted semen, a fluid (or "humor") which "has so great an influence upon the corporeal powers, and upon perfect digestion,...that the loss of an ounce of this humour would weaken more than that of forty ounces of blood." "Evidence" came from bizarre case histories: a soldier whose skull was found full of blood after he "died of an apoplexy, in the very act of coition," or a man "whose brain was so dried up [from masturbation], that it was heard to rattle in the pericranium"!

Perhaps we should not be too hard on Tissot. Many ancients held similar beliefs, as do some Taoist and Tantric Hindu religions today. And some of the physiological changes of sexual arousal—redistribution of blood in the body, muscular tension, and rhythmic contractions at orgasm—can in rare cases cause anything from a harmless (but intense and long-lasting) headache to a hemorrhage in the brain.

Sexual impulses could be reduced, Tissot believed, by a proper diet, such as mealy flour and milk. In the early 1900s, health reformers used such ideas to invent products we eat today, such as graham crackers and corn flakes. Presumably, few consumers today choose these foods in the hopes of reducing their urge to masturbate, or even know that corn flakes were invented by a man running a mental hospital who wanted to keep his patients from masturbating in the hallways!

No less an authority than the Boy Scout Handbook of 1925 sounded Tissot's alarm:

"In the body of every boy who has reached his teens, the Creator of the universe has sown...the most wonderful material in all the physical world. Some parts of it find their way into the blood, and through the blood gives tone to the muscles, power to the brain, and strength to the nerves. This fluid is the sex fluid...Any habit which a boy has that causes this fluid to be discharged from the body tends to weaken his strength, to make him less able to resist disease."

Distinguishing between semen and testosterone had to await the discovery of the hormone in the early 1930s. And though Tissot claimed that women, like men, would fall ill and go insane with excessive orgasms (in spite of the absence of any female fluid analogous to semen), the Girl Scout Handbook from the mid-1920s stayed mum on masturbation.

▶ COLD SHOWERS AS REMEDY FOR SEXUAL URGES?

students were taking classes in sex, and their own sex lives were no longer supervised by their schools. Gay students, inspired by the 1969 Stonewall riot in Greenwich Village, came "out of the closet" and began demanding recognition and respect.

And then came AIDS: Acquired Immune Deficiency Syndrome. By the time AIDS began its deadly attack sometime in the 1970s, America had already spent millions of dollars on basic biomedical studies, so researchers had something of a running start and were soon developing a number of drugs and therapies.

On the other hand, we had invested nearly nothing in sexology. Accordingly, this aspect of our response to AIDS limps along, as we struggle to design safer-sex campaigns with no basic research to guide us. Why do people enjoy what they fantasize sexually? What types of fantasies are common? Uncommon? What personality types are more or less likely to respond to a safer-sex campaign? At the moment, no one knows the answers to these simple, and vital, questions.

◄

IN THE VICTORIAN ERA, "GAY" WAS A SLANG TERM DESCRIBING PROSTITUTES.

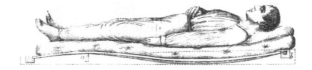

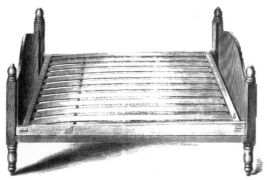

A BUSY NIGHT

While sleep behavior still looks as simple as it did 150 years ago, the scientific study of the brain has revealed much fascinating internal complexity. Discoveries that have been made include the existence of a regular cycle of brain activation in sleep that is associated with dreaming. It has also been recognized that sleep is not as safe as it looks but may impair such vital functions as respiration. And it has recently been demonstrated that sleep is not merely a restful luxury, but is essential to life: loss of sleep is invariably fatal if it persists for more than three weeks.

J. Allan Hobson   All of the important discoveries about sleep have occurred since 1928 when the German psychiatrist Adolf Berger reported that the electrical activity of the human brain could be recorded by attaching electrodes to the scalp and connecting them to an amplifier-recording system called the electroencephalogram, or EEG. One of the first things that Berger noticed was that his subjects' brain waves became slower and larger when they fell asleep.

Using the EEG, scientists were able to chart brain waves throughout entire nights of sleep; they found that periodic rises and falls occurred in the EEG pattern at regular 90-minute intervals. Each 90-minute cycle begins with large, slow EEG waves and deep, restful sleep; each cycle ends with small, fast EEG waves and light, dreaming sleep. The association of dreaming with EEG activation and with rapid-eye movement (or REM) sleep put the psychological study of dreams on a firm physiological footing for the first time. It soon became clear that dreams were more vivid, more fearful, and more strange when the eyes were moving rapidly in REM than when they were still.

Following the discovery of REM sleep and its relationship to dreaming by Eugene Aserinsky and Nathaniel Kleitman (US) in 1953, scientists were able to locate the control system in a small part of the stem of the brain called the pons, a region which also controls respiration and other automatic behaviors. Scientists now believe that the differences between waking and dreaming are the result of chemicals released when different groups of nerve cells in the pons change their discharge patterns.

With the help of the EEG, doctors have been able to record the sleep of patients who cannot easily fall asleep, or who fall asleep too easily, or who experience abnormal movements within their sleep. Many interesting sleep

◄

MARC CHAGALL FOUND
INSPIRATION FROM
JACOB'S DREAM OF
A HEAVENLY LADDER.

disorders have recently been brought to light; one, the tendency of some older overweight men to stop breathing while asleep, is quite unfortunate.

The severe impairment of breathing that afflicts these patients while they sleep is the result of an exaggeration of the normal tendency for body functions, including muscle and heart action, to slow down when we first fall asleep. Breathing then becomes deeper and may sometimes stop, or may be obstructed by snoring which may result in frequent awakenings and unrestful sleep. This problem, which is called the sleep apnea syndrome, can be alleviated by externally assisting the patient's breathing.

Scientists have recently found that, when consistently deprived of sleep, experimental animals cannot survive beyond three or four weeks. They die because they cannot maintain their body weight (despite being able to eat as much as they want) and because they cannot regulate their body temperature. If the animals are allowed to sleep, even if they are near death they quickly recover their body weight and temperature equilibrium.

These dramatic results indicate that sleep helps the brain to maintain an even flow of energy and heat throughout the body. It is as if the body's thermostat went into the shop each night for a preventive maintenance check. It is also clear that sleep, and especially REM sleep, serves to rebalance the forces of excitation and inhibition in the brain, making it possible for us to function optimally in our thinking, to maintain an even keel emotionally, and to avoid the excesses of extreme conditions like manic-depressive psychosis and epilepsy.

Modern scientific research on sleep has progressed hand-in-hand with our rapidly increasing understanding of the brain. It is now known, for example, that in order for us to fall asleep, an internal 24-hour clock in a part of the brain called the hypothalamus must function properly. The hypothalamus regulates body weight, sex-hormone release and temperature. In order for us to enter REM sleep, groups of nerve cells in another part of the brain, called the pons, must stop liberating chemicals that help us stay alert, while other nerve-cell groups liberate the chemicals that stimulate our brain to dream. As a consequence of these changes, nerve cells that regulate breathing may become less active, less responsive and more disorganized.

These dramatic shifts in the chemical balance of the brain during REM help us understand why our dream experience is so different from that of the waking state. The fact that brain-chemical shifts involved in triggering dreams are similar to those now known to occur in some forms of mental illness may account for the similarity of dreams to hallucinations and delusions of madness. This link has already led to more effective drug treatments of severe depression. Sleep research has become central to our understanding of how the brain mediates consciousness under normal and abnormal conditions.

Knowing that the brain's chemical balance is upset in sleep also tells us something about the risks and benefits of sleep. The risks, which include cessation of breathing leading to frequent arousals and subsequent fatigue, occur because the rules of brain regulation change with changes in brain chemistry. The benefits, which include our ability to be energetic and our ability to attend, to think and to remember, indicate that the brain also profits from its regular change in chemical balance.

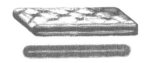

Fortunately, the benefits of sleep almost always outweigh the risks. And understanding how both are chemically mediated makes it possible for us to tip the brain's chemical balance in a favorable direction. The new science of sleep offers effective and safe approaches to the medical treatment of brain-chemical imbalance.

The Stranger in the Woodshed

The importance of sleep to the maintenance of healthy brain and mind functions was brought home to me with a vengeance when I suffered the dramatic effects of several successive nights of insomnia, one of them total. Whenever I travel to scientific meetings, I sleep poorly and when I took the "red-eye" night flight home from a five-day neuroscience convention I didn't sleep a wink. I was uncomfortably wedged in the center of a three-seat airplane row.

When I arrived in Boston, I drove with my family to our country place in Vermont. I decided to enjoy the lustrous beauty of the afternoon and go to bed early that night. But I was brought down long before dust when my brain suddenly went haywire.

I was in my woodshed stacking kindling when I was startled to discover a stranger standing behind me. Whipping my head round in a panic, I could clearly visualize the alarming intruder. The next thing I knew, I was lying on the ground some 15 feet away, having been hurled to the ground by some internal force, and having lost consciousness for a second or two. While it is impossible to be sure of exactly what happened, a realistic speculation is that my dream-like hallucination, my brief loss of consciousness and my fall were all caused by seizure-like activity in my REM-sleep-deprived pons. Fortunately, I was cured by a good night's sleep; I have never since had a recurrence of any hallucination.

THE VOYAGE OUT

A hundred and fifty years ago, steam was the cutting edge of technology. Flying was restricted to hot-air balloons. Heavier-than-air craft, to say nothing of space flight, were the stuff of dreams and fantasy. The other worlds in our solar system were known only by their indistinct and distorted images acquired by small telescopes peering through an ocean of air. Much that the best astronomers of the time thought they knew about the other planets, moons, comets, and asteroids we now know to be wrong. This remarkable change in our knowledge of our Solar

Carl Sagan System has been brought about only in the last few decades by interplanetary flight, chiefly by the United States and the former Soviet Union. Consider this example: Voyager 1 and Voyager 2 are the ships that opened the Solar System for exploration, trailblazing a path for future generations. Before their launch, in August and September 1977, we were almost wholly ignorant about most of the planets in our Solar System. In the next dozen years, these explorers provided our first detailed, close-up information about many new worlds—some of them known previously only as fuzzy disks in the eyepieces of ground-based telescopes, some merely as points of light, and others whose very existence was unsuspected. After nearly two decades, both spacecraft are still returning reams of data.

Voyager 1 and Voyager 2 have taught us about the wonders of other worlds, about the uniqueness and fragility of our own, about beginnings and ends. They have given us access to most of the Solar System—both in extent and in mass. They are the ships that first explored what may be the homelands of our remote descendants.

Today, U.S. launch vehicles are too feeble to get such spacecraft to Jupiter and beyond in only a few years by rocket propulsion alone. But if we're clever (and lucky), there's something else we can do: We can fly close to one world, and have its gravity fling us on to the next. This approach, called a gravity assist, costs almost nothing but ingenuity. It's something like grabbing hold of a post on a moving merry-go-round as it passes—to speed you up and fling you in some new direction. The spacecraft's acceleration is compensated by a deceleration in the planet's orbital motion around the sun, but because the planet is so massive compared to the spacecraft the planet slows down hardly at all. Each Voyager spacecraft picked up a velocity boost of nearly 40,000 miles per hour from Jupiter's gravity, and Jupiter in turn was slowed down in its motion around the

◄

SCIENTIFIC AMERICAN HAS CHRONICLED OUR ASCENT INTO THE STARS FROM A TRANSATLANTIC BALLOON CROSSING IN 1873 THROUGH THE VOYAGER 2 ENCOUNTER WITH SATURN AND BEYOND.

sun. By how much? Five billion years from now, when our Sun becomes a swollen red giant, Jupiter will be one millimeter short of where it would have been had Voyager not flown by it in the late twentieth century.

Voyager 2 took advantage of a rare lining-up of the planets: A close fly-by of Jupiter accelerated it on to Saturn, Saturn to Uranus, Uranus to Neptune, and Neptune to the stars. But you can't do this anytime you like: The previous opportunity for such a game of celestial billiards presented itself during the presidency of Thomas Jefferson. We were then only at the horseback, canoe, and sailing-ship stage of exploration. (Steamboats were the transforming new technology just around the corner.)

Since adequate funds for a more far-reaching mission were unavailable, the National Aeronautics and Space Administration's Jet Propulsion Laboratory (JPL) could afford to build spacecraft that would work reliably only as far as Saturn. Beyond that, all bets were off. However, because of the brilliance of the engineering design—and the fact that JPL engineers who radioed instructions up to the spacecraft got smarter faster than the spacecraft got stupid—both spacecraft were able to go on to explore Uranus and Neptune. These days, they are broadcasting back discoveries from well beyond the most distant known planet of the sun.

LUNAR LANDSCAPE AS IMAGINED IN 1874

We tend to hear much more about the treasures a ship brings back than about the ship itself or the shipwrights that built it. It has always been that way. Even those history books enamored of the voyages of Christopher Columbus do not tell us much about the builders of the Niña, the Pinta, and the Santa Maria, or about the principle of the three-masted caravel. Voyager 1 and Voyager 2, however—and their designers, builders, navigators, and controllers—are examples of what science and engineering, set free for well-defined peaceful purposes, can accomplish.

At each of the four giant planets—Jupiter, Saturn, Uranus, and Neptune—one or both spacecraft studied the planet itself, its rings, and its moons. At Jupiter, in 1979, they braved a dose of trapped charged particles a thousand times more intense than a human could survive; enveloped in all that radiation, they discovered the unexpected rings of the largest planet, the first active volcanos beyond Earth, and a possible underground ocean on an airless world—among a host of other surprising discoveries. At Saturn, in 1980 and 1981, the spacecraft survived a blizzard of ice and found not a few new rings, but thousands. They examined frozen moons mysteriously melted in the comparatively recent past, and a large world with a putative ocean of liquid hydrocarbons surmounted by clouds of organic matter.

On January 25, 1986, Voyager 2 entered the Uranus system and reported a procession of wonders. The encounter lasted only a few hours, but the data faithfully relayed back to Earth have revolutionized our knowledge of the aquamarine planet, its 15 moons, its pitch-black rings, and its belt of trapped high-energy charged particles. Three and a half years later, on August 25, 1989, Voyager 2 swept through the Neptune system and observed, dimly illuminated by the distant Sun, kaleidoscopic cloud patterns and a bizarre moon on which plumes of fine organic particles were being blown about by the astonishingly thin air. And in 1992, having flown beyond the outermost known planet, both Voyagers picked up radio emissions thought to emanate from the even more remote heliopause—the place where the wind from the sun gives way to the wind from the stars.

Because we're stuck on Earth, we're forced to peer at distant worlds through a blanket of distorting air. Much of the ultraviolet, infrared, and radio waves they reflect or emit do not penetrate our atmosphere. The Voyager spacecraft now ascend with stark clarity into the vacuum of space, returning four trillion bits of information to Earth, the equivalent of about 100,000 encyclopedia volumes.

▲

IMAGINATIVE
NINETEENTH-CENTURY
SPACE TRAVEL
ILLUSTRATED IN
SCIENTIFIC AMERICAN

When we add to this the accomplishments of NASA's Mariners, Vikings, Pioneers, Apollos, of the Magellan radar orbiter of Venus (which returned more data than all other robotic missions combined), of the Soviet Luna and Venera series of spacecraft, and of the European Space Agency's Giotto mission to Halley's comet, we can see the extraordinary character of our time. We have nearly completed the preliminary reconnaissance of our Solar System. We have visited more than 70 planets, moons, comets, and asteroids (plus one newly discovered moon of an asteroid). We have witnessed close-up all the planets known to the ancients, all the planets that circle the sun—with the sole exception of distant Pluto. We have gained a perspective on the uniqueness and fragility of our own world. When much of what occupies our front pages will have been consigned to dusty archives, our time will be remembered as the historic moment when we first learned what the rest of the Solar System is like, when our machines and we ourselves first touched other worlds.

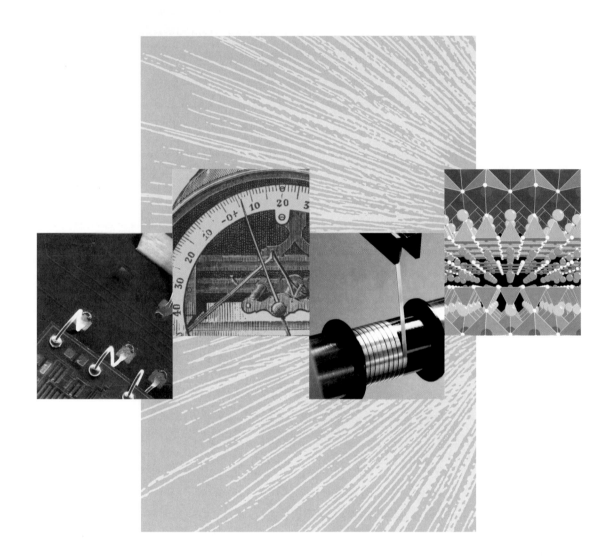

GOING WITH
THE FLOW

After my six-year-old son Miles saw a levitating skateboard in the movie *Back to the Future*, he asked me to make him one for his birthday. This was not really an unreasonable request for a child who had been watching me levitate small magnets in mid-air over superconducting ceramic pellets since just after he learned how to walk. Although I didn't make him the skateboard, such a mode of transport would not be very far beyond the reach of present-day technology: since the dramatic discovery of high-temperature superconductivity in 1986, we are moving closer to the day when such things might indeed be possible.

When electrical current travels through metal wiring in our homes and offices, a significant amount of the energy the electrons started out with is lost through collisions with atoms in the crystal lattice of the wire and is dissipated as heat.

Robert J. Cava
Superconductivity is the uncanny ability of some elements, metallic alloys and chemical compounds to conduct electricity with essentially zero electrical resistance. Superconducting materials have a normal degree of metallic electrical resistance at room temperature (comparable, for instance, to that of copper), but when cooled to sufficiently low temperatures, they suddenly undergo a transition to zero-resistance state. This "critical temperature," T_c, is generally close to absolute zero, limiting the use of superconductors in everyday technology. Superconductors potentially make great windings for high-field electromagnets; owing to their zero resistance, little heat is generated by the passage of the large electrical currents necessary to produce high fields. It is in such magnets, as for instance those found in modern MRI machines, that superconductors play an important role in everyday life.

Superconductors display an additional remarkable property, known as the Meissner effect. When a magnetic field is applied to any normal material, it penetrates to interact with the atoms and electrons present. When a superconductor is cooled in a magnetic field, it acts like a normal material above T_c. But below T_c, for small magnetic fields, the field is completely expelled from the superconductor. The manner in which the flux finally penetrates into the superconductor with increasing magnetic field strength is an important consideration for technological applications. An intermediate state in which the magnetic field is partly repelled and partly penetrates is what makes stable levitation possible.

◄

LEFT TO RIGHT:
SUPERCONDUCTORS
SUCH AS SILICON
CHIPS UTILIZING
COPPER OXIDES ARE
USEFUL AT EXTREME
TEMPERATURES;
SUPERCONDUCTING
WIRE CAN BE WOUND
INTO COILS TO MAKE
MAGNETS; AND
IF YOU LOOK AT A
SUPERCONDUCTOR
UNDER A MICROSCOPE,
YOU CAN SEE THE
COMPLEX STRUCTURE
OF ITS ATOMS.

Superconductivity was first discovered by H. Kammerlingh Onnes in Leiden in 1911 for cooled solid elemental mercury. About 30 of the pure crystalline elements have been found to be superconducting, ranging from the high Tc of niobium to the low Tc of tungsten. Elements are "type I" superconductors: the application of a magnetic field (such as would be self-generated by a magnet made of superconducting wire) abruptly destroys superconductivity above a "critical field" strength, where the field suddenly and completely penetrates the superconductor. The discovery that the critical fields for elemental superconductors are too low for them to be of practical use was a great disappointment, and relegated superconductors to the role of "puzzling physical phenomenon" for four decades after their discovery.

Two important advances occurred in the 1950s. Superconducting metallic alloys (metals made by chemically reacting two or more elements together) did not

abruptly become non-superconducting in applied magnetic fields, but instead lost superconductivity only gradually as the magnetic field increased. From these "type II" superconductors, high-field magnets could be fabricated, and now the potential for practical application could be explored. In addition, a successful theoretical explanation for superconductivity

THE DISORDERED GRAINS *(LEFT)* STRONGLY HAMPER THE FLOW OF ELECTRICITY, WHILE ALIGNED GRAINS *(RIGHT)* ALLOW UNIMPEDED ELECTRICAL FLOW.

was finally put forward, by John Bardeen, Leon Cooper and Robert Schrieffer, in 1957 (the "BCS Theory"). Although electrons should normally repel each other, under certain special conditions in solids, the theory said, they could "condense" into a coherently acting sea of loosely bound electron pairs which could not lose energy to the underlying crystal lattice and therefore experienced no resistance to their motion. For superconductors then being studied, the "glue" holding the pairs together was found to be a favorable coupling between the electron motion and the crystal's atomic vibrations, called "electron-phonon coupling."

In the 30-year period between the discovery of type II superconductivity and the mid-1980s, the science and technology of superconductivity progressed at a slow rate, with perhaps the most important milestone being the manufacture and widespread use of superconducting magnets in MRI machines and scientific instruments. Then, beginning in late 1986, a series of discoveries occurred which stunned the scientific world. It turned out that superconductivity occurred in ceramic materials (which are usually electrically insulating) based on copper oxides at unprecedentedly high temperatures. These new superconducting materials have set the world of physical science on its ear. Most agree that it is impossible for conventional coupling between electrons and phonons to be the "glue" that holds the superconducting charge carrier pairs together at such high temperatures. Consequently, a whole new theoretical explanation is needed. Many exotic ideas have been proposed, but as yet there is no consensus as to which might be correct. Also, with such high Tc's, immersion in relatively inexpensive and easy-to-handle liquid nitrogen at low temperatures seems to promise more widespread and inexpensive technological application. World-wide effort is now being expended to overcome the problems that the ceramic superconductors have in carrying high currents in high magnetic fields and in making long lengths of wire for electrical transmission purposes. There is also considerable interest in creating new superconductors as a potential part of the faster, more sophisticated communications and computing systems of the future.

The Woodstock of Physics

Although the first report of superconductivity in copper oxides in late 1986 was rather modest, at a critical temperature only five degrees higher than the previous metallic record-holder, the discovery of superconductivity at 92 degrees Kelvin (more than four times the old record) in the months that followed sent shock waves through the world of physical science. Superconductivity above 72 degrees Kelvin, the boiling point of liquid nitrogen, a common, inexpensive coolant, was considered by some to be the "Holy Grail" of solid state physics.

At the annual meeting of the American Physical Society in New York, which took place only months after the discoveries, excitement was at a fever pitch, almost to the point of hysteria. At
a hastily organized special session that began at 7 PM and lasted well into the early hours of the morning, researchers from around the world presented their latest results—often merely hours old, freshly faxed to their hotel by feverish co-workers back in the lab. Thousands attended the talks in a hotel ballroom designed to accommodate a small fraction of that number. In a New York Times article published the next day, the meeting was aptly called, "The Woodstock of Physics." Just as the Woodstock rock concert was the defining moment for a generation of young Americans nearly 20 years before, a generation of solid-state scientists was unified in the search for understanding and applying high temperature superconductors at the Woodstock of Physics in 1987.

BENEFITS AND RISKS

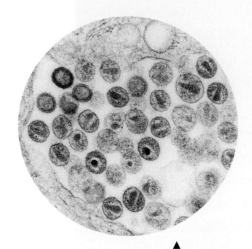

▲

HUMAN IMMUNO-
DEFICIENCY
VIRUS (HIV)

1981

THE U.S. CENTERS FOR DISEASE CONTROL

first recognizes acquired immunodeficiency syndrome (AIDS), a fatal disease that has since been shown to be caused by human immunodeficiency virus, or HIV. Transmitted primarily via blood products and semen, HIV destroys the human immune system, making its victims susceptible to opportunistic infections and secondary cancers. Despite more than a decade of intensive research, the disease continues to spread virtually unabated.

82

HUMAN INSULIN

becomes the first genetically engineered product.

84

DNA PROFILING,

a technique that isolates and types the unique DNA code found in blood, semen or hair left at a crime scene, is developed to confirm the identity of criminal suspects.

85

AT&T BEGINS PAVING

the information superhighway when it sends the equivalent of 300,000 simultaneous telephone conversations—or 200 high-resolution television shows—over a single optical fiber.

86

ENGINEERS AT THE CHERNOBYL

nuclear power plant in the northern Ukraine illegally disable the plant's emergency backup systems, perform an unauthorized test and accidentally initiate an uncontrolled chain reaction in a reactor core. The following day, an explosion rips the top off the containment building and expels eight tons of radioactive material into the atmosphere—triggering the worst nuclear disaster in history.

A NEW ERA begins at *Scientific American* when it is purchased by the Georg von Holtzbrinck Publishing Group.

88

THE U.S. PATENT AND TRADE OFFICE issues a patent to Harvard Medical School for a genetically engineered mouse. It is the first U.S. patent ever issued for a vertebrate.

87

AS MILLIONS OF TELEVISION VIEWERS watch in horror, the space shuttle Challenger explodes during its first minute of flight, killing all seven astronauts on board.

SWISS PHYSICIST KARL MUELLER and German physicist Johannes Bednorz develop a copper oxide ceramic that achieves super-conductivity, or complete loss of electrical resistance, at a relatively warm 30°C above absolute zero (-273°C)—fully 20°C warmer than previous superconductors. Their discovery holds out the promise of cheap, liquid-nitrogen cooled supercon-ductors for applications ranging from magnetically levitated trains to brain-mimicking supercomputers and earns Mueller and Bednorz a Nobel Prize the very same year.

89

CONCERNS ABOUT GLOBAL WARMING deepen after the previous summer's intense drought. Some scientists warn that continued fossil-fuel burn-ing and other industrial activities will cause carbon dioxide and other atmos-pheric greenhouse gases to double over the next 40 years—pushing up global temperatures 2°C to 5°C and prompting a polar-cap meltdown that will sub-merge entire coastal cities and ecosystems.

POLAR STRATOSPHERIC CLOUDS SUCH AS THESE HELP TO INITIATE CHEMICAL REACTIONS THAT DESTROY OZONE.

▼

BRITISH CHEMIST Martin Fleischmann and U.S. chemist B. Stanley Pons create a storm of controversy when they report they have observed nuclear fusion take place at room temperature in a common electrolytic cell. If substantiated, their claim will revolutionize the global production of cheap energy, but repeated attempts by other labora-tories to duplicate their results prove inconclusive or unsuccessful.

THE COLD WAR ENDS with the falling of the Berlin wall, the reunifica-tion of Germany and the impending dissolution of the U.S.S.R.

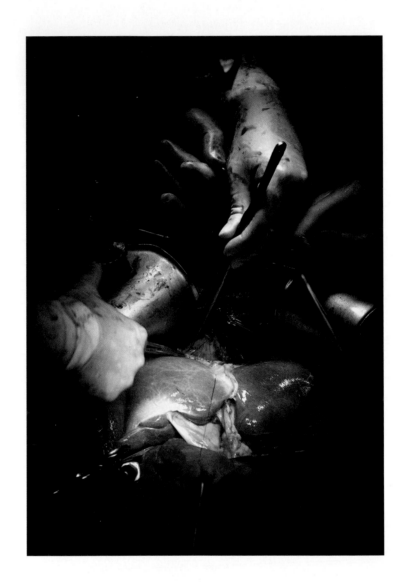

TRIUMPHS IN
THE SURGICAL THEATER

A century and a half ago, surgical procedures were done without anesthesia, and a major step forward was taken with the discovery of ether as the first anesthetic in 1846. This was a tremendous boon for all concerned—the patient no longer felt pain during the operation, and the surgeon could now operate less hurriedly and with less apprehension. Describing the anxiety and concern of the surgeon performing operations upon patients who were awake, Valentine Mott, the famous professor of surgery at the College of Physicians and Surgeons at Columbia University, said, "Operating in some deep, dark wound along the course of some great vein, with thin walls,...how often I have dreaded that some unfortunate struggle of the patient would deviate the knife a little from its normal course, and that I, who fain would be the deliverer, would involuntarily become the executioner, seeing my patient perish from hemorrhage which is one of the most appalling forms of death." These worries and risks were virtually eliminated with the advent of anesthesia; clearly, it is one of the most important medical discoveries of all time.

**David
C.
Sabiston, Jr.**

Joseph Lister, in Scotland, was the first to recognize the importance of germs as a cause of infections following surgical operations. To prevent infection, he introduced antiseptic surgery. He developed and implemented techniques designed to kill germs on the skin as well as those on the surgical instruments and on drapes surrounding the patient, all of which were sterilized before the operations began. This precaution has saved many lives and is now uniformly taken throughout the world in all surgical operations.

Scientific surgery saw another major achievement when Theodor Kocher of Switzerland removed the thyroid gland for goiter in more than 5,000 patients with an exceedingly low death rate, thus showing that surgery could be made quite safe. For his fine work on the surgical removal of goiter and also for his recognition that secretions of the thyroid gland are essential for a normal life and must be replaced by thyroid extract from animals, Kocher became the first surgeon to win the Nobel Prize in Medicine in 1908.

It is interesting to note that, prior to 1900, it was not possible to repair a blood vessel, whether an artery or a vein, after it had been injured or divided. When he was a young medical student in 1894 Alexis Carrel was deeply shaken by the assassination of the President of France by an anarchist who stabbed him in the abdomen. The weapon caused a serious injury to a major vein in the liver, and the president died shortly thereafter from massive hemorrhage. Over the course of many experiments, Carrel eventually perfected his technique for joining blood vessels with stitches; the vessels healed successfully and regained normal function. He also transplanted organs such as the kidney. This enormous advance allowed surgeons to perform many new operations which had previously not been possible. The new technique also prevented many deaths from injuries to blood vessels by controlling hemorrhage. The importance of this contribution by Carrel was widely noted, and he was the second surgeon to receive a Nobel Prize in Medicine, in 1912.

Until 1922, diabetes was a dreaded disease, as it caused so many medical problems and usually meant a premature death for its victims. Patients with diabetes who underwent surgery often suffered many post-operative complications because they were unable to properly utilize glucose, the basic blood sugar. An

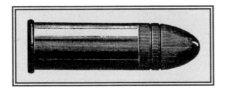

▲

BEFORE THE DISCOV-
ERY OF ANESTHETICS,
MANY A PATIENT HAD
TO "BITE THE BULLET."

orthopedic surgeon in Canada, Frederick Banting, was the first to demonstrate that the disease was caused by a deficiency of insulin, which is produced by the pancreas. He obtained insulin from the pancreas of animals which, when injected by needle under the skin, corrected sugar metabolism in diabetic patients. This advance greatly improved the health of diabetics, and significantly extended their lives.

Until the late 1930s, physicians sought in vain to understand why injured patients experienced falling blood pressure followed by failure of a number of body organs. This condition, called shock, was generally thought to be caused by a toxic substance generated by injured tissues. After much research, Alfred Blalock was able to show convincingly that at the site of a wound, whether resulting from a gunshot, automobile accident, or other cause, a leakage of fluid from the blood stream at the injured site into the tissues induced both a low blood pressure and organ failure of varying degrees. Replacing this fluid with plasma or blood or even salt solutions usually caused blood pressure to return to normal, and restored bodily functions. This discovery saved many lives and was particularly important to the military in combat. An eminent scientist from Oxford University, Sir James Pickering, said of this work, "The conclusion that emerged from World War I was that shock was a traumatic toxemia produced by the effects of substances released from injured muscles. It needed the genius of Alfred Blalock and the experience of the Second World War to show that this was not so." Blalock's basic contributions to the understanding of the causes of shock have since been responsible for saving thousands of lives among injured patients throughout the world.

The introduction of antibiotics was another advance that made surgical operations more successful. The British bacteriologist, Sir Alexander Fleming, made the observation in 1928 that a fungus, penicillin, could kill dangerous germs. Later, Sir Howard Florey and Ernst Boris Chain were able to purify the product of the fungus and to concentrate it in large amounts in a form which could be given to patients with serious infections. Today, penicillin is known as a "magic" drug that has saved thousands of lives since World War II, when it was first administered to injured soldiers with life-threatening infections. It greatly reduced the death toll from war wounds and rapidly began to be used to treat civilian patients. This antibiotic, and many others like it which have since been developed, are absolutely essential to the practice of modern medicine.

Open-heart surgery was first made possible by the invention of the heart-lung machine. The device temporarily replaces the heart in pumping blood throughout the body, and also substitutes for the lungs in providing oxygen to the blood. The introduction of this machine was a painstaking project; it was many years before the technology could be successfully applied. It was John F. Gibbon and his wife, Mary Hopsinson Gibbon, of Philadelphia, who pioneered this major advance in medical research. 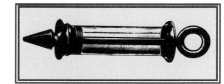 For twenty-two years, they devoted themselves to making the project a success. Although many colleagues tried to discourage them, claiming that such a device could never work, they pursued their dream year after year. With the help of engineers from IBM, they adjusted their machine many times until it was ultimately perfected. Their perseverance, hard work and faith were finally rewarded. In 1953 Gibbon, himself a heart surgeon, connected the heart-lung machine to a 17-year-old girl with severe heart failure; he then opened her heart and patched a large hole inside the organ. She rapidly improved once the heart failure was alleviated and from that time on, thousands of infants, children, and adults have had a variety of heart defects successfully corrected using the heart-lung machine.

Artificial heart valves are another significant step in the advance of heart surgery. By 1960, Albert Starr and others had developed plastic valves that could successfully replace diseased heart valves, rendering seriously ill patients essentially normal again. Later, tissue grafts were introduced which further enhanced heart surgery.

By 1962, the bypass graft, developed by the author, was designed for patients suffering from coronary artery disease, which causes heart attacks.

In this operation, certain veins are removed, usually from the patient's leg, and are joined to the arteries of the heart. Currently each year, more than 300,000 patients in the United States with coronary artery disease are treated with coronary artery bypass grafts. Many more of these procedures are performed around the world with a high rate of success both in relieving the patient's pain and in significantly increasing their life expectancy. Ultimately, it is hoped that the basic cause of blocked arteries and heart attacks will be discovered so that they can be prevented altogether. Until that time, bypass grafts will continue to be an important treatment.

The most recent advance in cardiac care has been a mechanical device that can be attached to the heart of patients who are in severe heart failure and are in need of a heart transplant. These patients are very likely to die if a donor heart is not immediately available. Since such hearts are scarce, it may be essential to attach the artificial heart to the body to keep the patient alive until a heart can be donated for transplanting. Artificial hearts have been used successfully for periods ranging from weeks to months, providing the precious extra time that makes the difference between life and death.

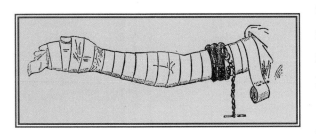

▲

THESE IMAGES, FROM
SCIENTIFIC AMERICAN
IN 1874, ILLUSTRATE
SURGICAL OPERATIONS
WITHOUT PROVOKING
THE USUAL HEMOR-
RHAGE BY USE OF A
26" ELASTIC BANDAGE
AND A TUBE OF
VULCANIZED RUBBER.

In 1954 Joseph E. Murray, in Boston, first successfully transplanted a kidney from one identical twin to the other. It was the first instance of successful transplantation of a vital whole organ, and it ultimately won Murray the Nobel Prize in 1990. However, although identical twins can receive tissues and organs from each other that will survive and function normally, an organ transplanted from anyone except an identical twin is usually rejected by the body and so fails and dies. It has been necessary to develop drugs to help prevent this rejection. Such a series of drugs was first discovered by George H. Hitchings and Gertrude Elion (US), and similar ones have made possible the successful transplanting of organs between patients who are not identical twins. For their outstanding contributions, Hitchings and Elion jointly received the Nobel Prize in 1988, many years after the significance of their discoveries were first recognized. Today, many organs are successfully transplanted, including the kidneys, pancreas, liver, heart, lungs, bone marrow and others, allowing the patient to live many additional years. Since human organs are scarce and not nearly enough are available for patients who need them, much work is currently being done to develop specially bred animals from whom organs can be transplanted into humans, using appropriate drugs to prevent rejection. Many believe that this kind of transplanting, called xenotransplantation, will become feasible within a few years. Several research teams are busily engaged in the project at present, and predict a successful outcome.

The treatment of cancer has long been a challenge to the medical profession. Some cancers can be removed from the body by surgery with complete success, but others cannot. Even cancers which can be removed surgically may recur in other parts of the body and cause the death of the patient. Charles B. Huggins, a urologic surgeon, showed in 1940 that certain hormones can be employed to fight cancer, especially prostate cancer, and this discovery has been responsible for adding many years of life to patients with this disease. For this work, Huggins received the Nobel Prize in 1966. Today, chemotherapeutic drugs are available that can kill cancers. For example, a cancer called chorioepithelioma, which sometimes occurs in the womb following a pregnancy, was formerly a dreaded problem, since it often spread to other parts of the body and caused death. It is now possible to administer a specific agent, called methotrexate, to patients with this cancer; nearly all survive

and most are cured completely. It is believed that gene therapy will play a substantial role in the treatment of cancers in the future. This is particularly true of cancers which have spread to other organs, and encouraging research currently in progress has engendered optimism that this new therapy will become feasible.

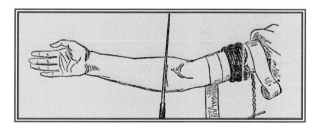

Already there are other diseases which respond to this form of treatment, and there is considerable optimism concerning its success in the management of cancer.

Quite recently laparoscopic surgery has become widely used in a number of operations. In this procedure, a small tube with a light attached to it is passed through a tiny (half-inch) incision in the abdomen; other surgical instruments can be inserted through even smaller incisions. Many major abdominal operations can be performed in this manner. Formerly, these operations required a standard surgical incision six or more inches in length, which could cause the patient considerable pain and discomfort for weeks afterward. With the laparoscopic approach, there is much less pain, and patients recover more rapidly and return to normal life more quickly. Moreover, they are in the hospital for much shorter periods and often go home either on the day of the operation or the following day. The same approach can be used for chest operations; this is called thoracoscopic surgery. Less time in the hospital, a more rapid recovery, and an earlier return to work result in a variety of economic savings which are of considerable significance today in reducing the cost of health care.

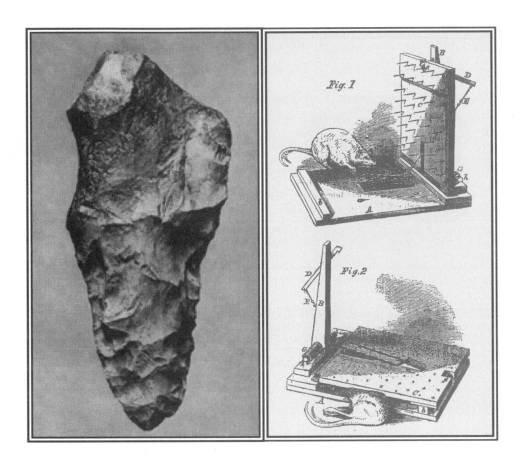

IMAGINATIVE
USE OF MACHINES

Scientific American and the science-based industrial revolution in America were born at about the same time. In 1837, the University of Alabama began teaching engineering to help the great cotton plantations of the South build an infrastructure, bringing an industrial dimension to agriculture. A quarter-century before the slaves were liberated by Abraham Lincoln, steam-powered technology had begun to diminish the economic value of slavery, just as Eli Whitney's invention of the cotton gin in 1793 had enhanced it. Technology (which has no moral conscience) has been changing the world ever since.

Lewis M. Branscomb

Technology—the imaginative use of machines and processes to solve certain physical problems—is older than recorded history. No one knows who first used a razor-edged piece of obsidian stone as a knife, or who figured out how to start a fire by friction of wood on wood. Did an imaginative cave-woman invent the first loom? Technologies that enriched the lives of early humans were being invented for thousands of years with no help from science. The hairpin, paper clip and mouse trap were creations of clever inventors, not off-shoots of scientific discovery.

Nevertheless, progress in technology was agonizingly slow until modern science emerged in the eighteenth century. A wave of new technologies emerged in the early nineteenth century, driven by the winds of European science. Embodied in the language of mathematics, the new science broke through the barriers to invention.

Steam-driven ships, first proved out on rivers and inland waters, brought southern cotton down-river to ships bound for the textile mills of Massachusetts. Soon steam power drove the great sailing ships on the high seas. Both in the South and in New England it became clear that the ability to use science to create technology was a skill many had to learn. Engineering education, the province of West Point since the founding of the nation, became "civil" and evolved into the profession underlying all modern economies. In 1835, only nine years before *Scientific American*'s birth, the first four American degrees in "civil" engineering were awarded by the Rensselaer School (later R.P.I.) in Troy, New York.

The Massachusetts Institute of Technology, perhaps the world's most renowned fount of technological education and creativity, was chartered in 1861. But in the mid-nineteenth century, technology was already taking hold all over the young United States. Indeed, the South and the Northeast enjoy a curious

◀

A HAND-AX, FOUND IN FRANCE, IS APPROXIMATELY 200,000 YEARS OLD — ONE OF THE EARLIEST TECHNOLOGICAL INNOVATIONS. ALSO PICTURED IS AN 1860 ATTEMPT TO BUILD A BETTER MOUSETRAP.

coupling in their histories. The founder of MIT, William Barton Rogers, left his professorship at the University of Virginia to set up a comprehensive school of useful science in Boston because, according to O. Allan Gianinny, Jr., he believed that the "democratic and scientific values embodied in engineering education could not be fulfilled in a slave economy."

The discovery of gold and silver in the West created a keen demand for energy technology 100 years ago. In England, James Michael Faraday's scientific discoveries in electricity laid the scientific foundation for the next major energy technology to be developed after steam. Cornell University had created the first Electrical Engineering Department in the United States in 1888. Nikola Tesla, who sold his inventions on alternating-current motors to George Westinghouse, came to America in 1884. In 1890,

▶

ENERGY SOURCES, SUCH AS STEAM AND GASOLINE, FUELED THE FIRST WAVE OF TECHNOLOGICAL BREAKTHROUGHS. PICTURED HERE ARE A STEAM-POWERED PLOW FROM 1870 (TOP), A GAS-POWERED TRACTOR (BOTTOM) . . .

a Telluride, Colorado mine owner, named L.L. Nunn, sought the help of Westinghouse and Tesla to build an alternating-current power plant for his mine. With the help of 30 students brought to Telluride from Cornell, they provided this tiny mining town in the Uncompahgre Mountains of Colorado with the first central electric power in America.

Fourteen years after *Scientific American* began publishing, the first oil well was drilled in Titusville, Pennsylvania. Petroleum soon made the whale-oil lamp obsolete and replaced the horse with the automobile. Another energy technology revolution was unleashed. Moreover, abundant hydrocarbons gave rise to the entire structure of modern organic chemistry.

Today, technologies are multiplied not only by invention, but also by combining existing technologies. The Wright brothers' airplane combined the light-weight structure of the bicycle with the automobile's internal combustion engine. Their radical new creation was the airfoil, which was tested in a wind tunnel they had invented. Early in the twentieth century the industrial-research laboratory appeared and institutionalized invention-on-demand. Even since, new technology is created whenever needs arise and money and talent are made available to address them.

Most of us today take technology for granted, even as we enjoy the convenience it provides. How many electric motors are in a typical American home? If you look you will probably find more than 40. How many computer chips—those tiny electrical engines that do so much thinking for us—are in your home? More every day. Technology simplifies domestic tasks, allowing homemakers to also work outside the home, doubling the U.S. labor force.

Geography no longer defines community; communications and computer technology have created the "global village." People with common interests can now work together without having to live in the same place. The global information infrastructure—a planet-circling spiderweb of digital communications connecting tens of millions of computers—gives people political power that transcends national boundaries. Indeed, many credit the copying machine, the tape cassette, and the personal computer with destabilizing communism in the Soviet Union. Through television we share the thrills of the Lillehammer Olympics and the agonies of Rwandan refugees. In a few hours, one can jet to Norway to cheer the victorious, or to Rwanda to comfort the dying.

Scientific American surfed into the lives of American readers on the first of two great waves of new technology. That first wave was spawned by new sources of energy: steam, electricity, and gasoline, and more recently, nuclear and solar power. Energy technologies made possible enormous productivity and mobility. No longer were people limited by their physical strength. Motors magnified muscles a thousand fold.

The next wave broke on our shores only 40 years ago—the electronic computer, a motor for our minds. All the information technologies—telephones, TV, radio, satellites, personal computers, cellular telephones, and the software that instructs them to do our bidding—can be assembled into networks to bring all of human knowledge to our fingertips. More important, information can substitute for energy in many situations.

The advance in technology is, of course, closely bound up with humanity's relationship with the natural worlds. The first stage of technology emerged when people used nature more or less as they found it to make a spear, stone knife, or a deer skin cloak, to feed captured wild animals or to plant wild seeds. In the second

◄

. . . A 1902 STEAM CARRIAGE *(TOP)* AND A GAS-POWERED CAR *(BOTTOM)*.

The Computer

Some 40 million American homes have computers in them today, but computer technology has a longer history than many users may realize. A design for a digital computer was completed by Charles Babbage in 1837. His colleague, the mathematician Ada Byron King, Countess of Lovelace, pioneered ideas in computer software. An advanced computer language, ADA, is named for her. But Babbage's machine was never completed; the switching technologies needed were not yet developed. Indeed, the record of his work was lost for 100 years.

By the 1930s, mechanical calculators using punched cards were developed by IBM to help accountants manipulate their numbers. When Massachusetts introduced standardized tests in its public schools in the 1930s, it became apparent that there was no practical way to score so many test papers. Thomas J. Watson, chairman of IBM, donated a truckload of punch-card equipment to Columbia University, and the development of what became the programmable computer was on its way.

Key technical developments were made by Herman Hollerith, who built a machine for recording and analyzing the data from the U.S. Decennial Census. Then came World War II, and Presper Eckert, Robert Mauchly, and John Von Neumann invented programmable digital computers for solving mathematical problems in field artillery using mechanical switches developed for telephone exchanges. With the invention of the transistor at Bell Telephone Laboratories, the computer industry took off.

Ever since, the cost of purchasing a fixed amount of computing power has dropped by an astounding 15 percent a year, and it continues to fall. Whenever a reduction in price creates greater demand for a product, technology responds to supply it.

TABULATING AND ADDING MACHINE DESIGNED IN THE EARLY TWENTIETH CENTURY

stage, beginning about 10,000 B.C., people learned to adapt nature to their needs, using, for example, rubber-tree sap for tires, cotton fibers for clothing, lumber for houses, flints for arrow points, and the selective breeding of plants and animals for food and other products. The third stage, which brings us to mid-twentieth century, is the result of the industrial revolution. Here, people reconstruct nature by processing natural materials to create new materials and tools that don't exist in nature—concrete, fiberglass, synthetic fibers, glass, steel and paper.

In the fourth stage, which is now upon us, we are actually creating new forms of nature to our own specifications. Plants and animals that never existed come to be through genetic engineering, while we make materials unknown in nature through molecular redesign and build machines that make decisions faster than we can think.

Scientists and engineers create bacteria that eat spilled oil, lubricants that withstand higher temperatures than nature's oils, tomatoes that don't rot or bruise easily and plastics that degrade to harmless substances when we want them to.

◄

INFORMATION
TECHNOLOGY WAS
THE SECOND WAVE
OF GREAT BREAK-
THROUGHS.

Even as technology improves society by meeting our most urgent needs, it creates new problems we have never confronted before. When we discard what is not useful, including the waste products of new industrial processes, we create bad air, dirty water, and mountains of solid waste. These problems, in turn, can be addressed using still newer technologies. "Industrial ecology" analyzes the flow of energy and materials in industry in order to minimize waste and reuse as much as possible.

In order to know what technologies to develop, we have to consider our future and not our past needs. Technologies do not have lives of their own; they are a reflection of the culture and values of the societies that create them.

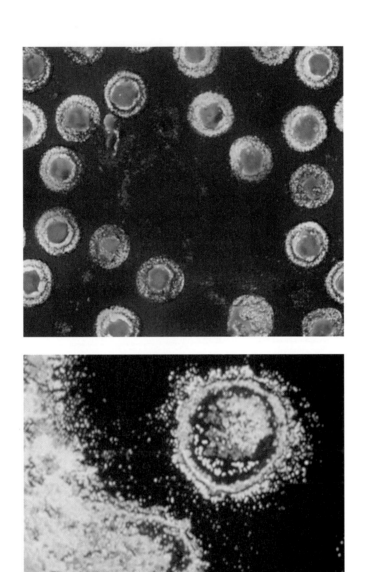

BENEATH THE WORLD OF APPEARANCES

The virus as an idea—an invisible layer of noxious substance beneath the world of appearances where "little animals multiply which the eye cannot see enter the body...and may cause grave disease"—is ancient. The Roman writer Marcus Varro made this connection over two thousand years ago, but it was not until the

**Robert C. Gallo
and
Howard Z. Streicher**

mid-nineteenth century that such an entity was seen in the real world. Far-reaching advances in biology inevitably wait upon major technological break-throughs. Leeuwenhoek's lens was powerful enough to see most kinds of microbes but not viruses. It wasn't until two centuries later that Koch and Pasteur discovered the major causes of infectious diseases and the principles of vaccination.

Quite early in the history of bacteriology research Beijerinck in Holland found that a disease of plants, tobacco mosaic virus, could be caused by an infectious agent much smaller than a bacterium. Frosch and Dahlem found the same was true for an animal virus, foot and mouth disease.

◄

MICROPORTRAITS
OF VIRAL KILLERS:
HEPATITIS B *(TOP)*
AND HIV *(BOTTOM)*

In general, we can picture a virus as a microorganism, often responsible for disease, which is capable of growing only in a living cell of a vulnerable host. Not every virus is harmful to its host. Some viral effects, such as the streaking of tulips, seem desirable, and today modified viruses are being used to deliver genes for a variety of purposes, including the treatment of some diseases. Over the past 50 years a parade of viruses has marched through the pages of *Scientific American*: herpes, polio, enteroviruses, influenza, cold viruses, rubella, EBV, rabies, hepatitis, animal leukemia viruses and finally human tumor-producing viruses like human T-cell leukemia virus and, most recently, the AIDS virus.

The history of virology may be divided roughly into three eras. The first, from 1870 to 1935, was when viruses were known mostly from the diseases and pathologic effects they induced. The second began around 1935 when the crystal-lization and visualization of tobacco mosaic virus ushered in an era of discovery into their physical nature. At the same time, laboratory culture systems enabled researchers to study viruses in the very cells they infect without having to deal with the whole organism. Recent research began in the early 1950s with the birth of molecular biology, and viruses have played an integral part in advances made in the field. Indeed, no area of biology has been more fruitful or more closely integrated with expanding knowledge of molecular biology. For example, viruses may transfer information, and were early exploited to study the way nucleic acids (the genetic material) work in cells, notably in the use of the phage system (viruses that infect bacteria) by Max Delbruck of the US.

Virus classification is based on the nucleic acid content of the virus and its structural proteins. Viruses may have genomes of DNA or RNA and be single- or double-stranded. These traits and the size and complexity of a virus govern how it replicates. On the outside, each virus species is ingeniously assembled from just a few proteins which the electron microscope reveals to be arranged in repeating and symmetric units. While it seems to be on the very edge of a living entity, in order to reproduce itself, the virus sets in operation a highly organized molecular factory within the cell it infects, often employing biochemical and immunological tricks to this end. Viruses have provided us with profound examples of the extent to which incredibly small particles can cause devastating effects through the information contained in their genes.

Where do new viruses come from? As Manfred Eigen of Germany has pointed out, viruses may change so fast that they exist as "quasi-species" which could further evolve into new viruses. Many suspect that most viruses originate from the DNA of a cell. Once a set of genes develops a way of reproducing itself and harnesses a polymerase to replicate itself, it may be on its way. The other way a new virus may come on the scene is if a virus that has been present at a low level in a species suddenly emerges when the population of the species is altered as in environmental change. It is almost certain that plagues like small pox, yellow fever, influenza and AIDS have all emerged during periods of new global contacts among people.

In 1911, Peyton Rous (US) isolated a virus from a chicken with sarcoma, a cancer of muscle tissue. In an era when most disease, but not cancer, was attributed to microbes, he showed that this agent could reproduce cancer when transmitted to another animal. Fifty years later, Rous was awarded the Nobel prize for his work. The case supporting viral causes of cancer was subsequently greatly advanced by many others, notably John J. Bittner of the U.S. (mouse mammary tumor virus), Ludwig Gross of the US (murine leukemia virus), and William Jarrett of Scotland and Max Essex of the US (feline leukemia virus). A monumental breakthrough was made through the remarkable conceptual insight of the American scientist Howard Temin and independently by MIT's David Baltimore of the discovery of the enzyme reverse transcriptase in 1970. This enzyme catalyzes the synthesis of a double-stranded DNA molecule from the single-stranded RNA of the virus. This process reverses the "usual" flow of genetic information, which at the time was deemed always to flow from DNA to RNA. Temin's hypothesis suggested that the RNA virus could become DNA and remain permanently in the host-cell genes. These discoveries opened several new lines of investigation, one of which led to the discovery of onc genes, fundamental to our understanding of all types of tumors. The century-long search for a human tumor virus resulted in the detection and discovery of the first human retrovirus by the end of the 1970s.

Almost at the same time, evidence was steadily emerging that several more common and well-known human viruses could also cause cancer. A herpes virus, EBV, was linked to the cause of African Burkitt's lymphoma, a childhood cancer. Because it is ubiquitous, the virus-associated tumor occurs only in particular areas. The virus is but one of several environmental factors needed to cause these cancers. It also is associated with lymphoma in immunodeficient patients. There is evidence that a similar role is played by certain other DNA viruses. Particular strains of papilloma virus (wart virus) are associated with genital and cervical cancers and hepatitis B and C viruses are linked to hepatoma (liver cancer). A second human retrovirus (HTLV-II) was also discovered at about the same time, but its role in human disease remains uncertain.

In the early 1980s a new epidemic of immunodeficiency (AIDS) was first recognized. Based on the cumulative discoveries of the previous 70 years, it was possible to isolate, mass produce, characterize and devise a blood test for a new group of human retroviruses in an exceptionally short time. The virus, now known as HIV-1, was the third human retrovirus to be discovered.

Generally cheap, safe and effective, vaccines have been used to control polio and to eliminate small pox. With HIV infection we have run into the formidable problem of rapid virus variation. The virus changes faster than the ability of our immune system to recognize and control it.

Antiviral therapy attempts to exploit subtle differences between the virus and the host. Some success with drugs, such as acyclovir for herpes viruses and AZT for HIV, have given us reason to be hopeful. Yet we know that viruses have evolved with us and continue to exploit the innermost workings of our biologic systems. "New" viruses such as HIV, Ebola fever virus, and hantavirus will undoubtedly continue to emerge in conjunction with social and ecologic changes. As we learn about infectious information in the form of prions (only protein) and viroids (only RNA), we may be taking a leap into another era of virology.

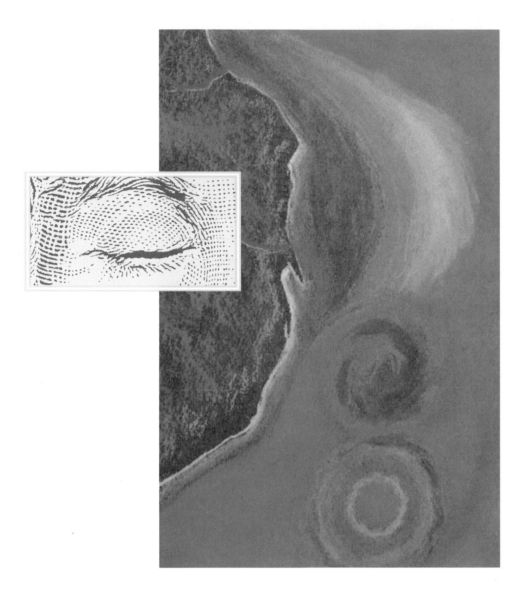

SIGHT SEEN

In 1860, the German psychologist Gustav T. Fechner inaugurated a new scientific discipline with the publication of his landmark text, *Elements of Psychophysics.* Psychophysics is the study of human perceptual capabilities, and, according to Fechner, may be carved into two domains. "Outer psy-

Torsten Wiesel

chophysics" deals with the relationship between a physical stimulus—a particular pattern of light, for example—and the sensation this light-pattern produces in the mind. "Inner psychophysics" has a more difficult aim—it describes the relationship between this visual sensation and the pattern of nerve-cell impulses within the retina and brain, that is elicited by light.

In the 1930s, the beginnings of a true "inner psychophysics" became possible when physiologists introduced techniques for recording light responses of single visual cells. Today, the most exciting aspect of vision research is the increasing interaction between inner psychophysics—now called neurophysiology—and outer psychophysics pioneered by Fechner and his contemporaries. Yet it was in the middle of the nineteenth century that researchers began attempting to correlate visual sensations with the activity of specific sets of cells in the visual system.

In 1825, the Czech physiologist Jan Purkynje, walking home in the dark hour before dawn after working all night in his laboratory, came upon a surprising observation. Purkynje found that he was unable to perceive the colors of flowers and other objects in the first dim light. Everything appeared to him in shades of gray. Yet there was a curious shift in the relative brightness of these gray objects according to the color of their surfaces. While red flowers appear not only colored, but relatively brighter than blue flowers during the day, the reverse was true to Purkynje's eyes in the dim light before dawn.

The German anatomist Max Schultze in 1866 proposed the first correct cellular explanation of the now-famous "Purkynje effect." Schultze concluded that the retina contains two distinct mechanisms for sensing light: one operates mainly at night, the other during the day.

Night vision is a function of the rod cells of the retina—photoreceptors specifically adapted to detect dim illumination. Rod vision is color-blind and resolves forms poorly. Rods are also maximally sensitive to light from the blue-green region of the spectrum, which accounts for the Purkynje effect. Day

◀

NEITHER THE WEALTH OF COLOR NOR SOME OF THE COMPLEX SURFACE PATTERNS ON THE EARTH CAN BE RE-CREATED BY TODAY'S TECHNOLOGY. UNAIDED HUMAN VISION CAN BETTER PORTRAY WHAT IS SEEN FROM SPACE AS SHOWN IN THIS DRAWING OF THE SOUTHERN TIP OF MADAGASCAR BASED ON OBSERVATIONS MADE WHILE IN ORBIT.

vision, on the other hand, is supported by retinal cone cells, which allow us to see both color and fine forms.

Schultze's studies raised three major questions, whose pursuit may serve to illustrate the growth of visual science as a whole over the last 150 years. What accounts for the rod's extreme sensitivity at night? How do nerve signals emanating from the cone cells allow us to see color? And how do we perceive the forms of objects around us? Research into these questions began as psychophysical studies of human visual performance but recently the cellular and even molecular bases of light sensitivity, color vision and form perception have begun to be revealed.

In 1942, the American biologist Selig Hecht demonstrated that individual rod cells operate at the physical limits of perception. A rod photoreceptor can signal the absorption of a single quantum of light. In 1979, Dennis Baylor at Stanford

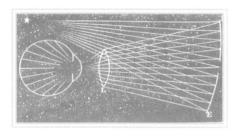

University and his colleagues at Stanford University succeeded in directly recording the electrical response of single rod cells to brief flashes of light, thus confirming Hecht's finding. Hecht's student, the American biochemist George Wald, turned to the study of light-sensitive rod pigment, rhodopsin. Recently, Lubert Stryer, a biochemist at Stanford, and others have shown how light absorption by molecules of rhodopsin can be transduced and amplified into a rod-cell signal by an intricate molecular signalling pathway. Today, we know a great deal about how sensitive our eyes are to the tiniest pulses of light.

In color vision, work by the nineteenth century physicists Thomas Young, Hermann Helmholtz, and James C. Maxwell led to the trichromatic theory of color perception. The trichromatic theory held that color perception resulted from the combined response of three different color-sensitive receptors in the retina. Beginning in the 1950s, William Rushton, Edward MacNichol, George Wald and their collaborators in the United States and Great Britain showed that the retina actually possesses three types of cones with three different color-sensitive pigments. More recently, the DNA encoding the color pigments has been isolated by Jeremy Nathans of John Hopkins School of Medicine, revealing the molecular basis of our ability to see the world in color.

Yet color perception only commences with cones. Studies by the nineteenth century German psychophysicist Ewald Hering pointed to a process by which distinct channels within the visual system compared the relative amount of color emanating from points of the visual scene. Our visual system's comparison of the relative intensity of red and green, and (separately) of blue and yellow, is the reason why the members of these color pairs appear opposite or "complementary" to each other. In the late 1950 and early 1960s, the American physiologist Russell DeValois, discovered cells in the monkey lateral geniculate nucleus (the relay station between the eye and the visual cortex of the brain) that have opposing properties. A cell may be excited by red light and inhibited by green light or vice versa. Another type of opposing color cell shows opposite responses to blue and yellow. The neural basis of Hering's theory of color perception is beginning to be revealed.

Perceiving the form of a figure before us depends on detecting the figure's contours and edges, and perceptually segregating the outline of the figure from a potentially confusing background. Think of the difficulty in perceiving the continuous contours of a tiger's head and flanks lying camouflaged in jungle underbrush. The Gestalt psychophysicists Max Wertheimer and Kurt Koffka showed in the 1920s that such figure-ground segregation is a global process that involves the interaction of contour-elements detected across the visual field. Today, we are still attempting to discover the neural basis of such Gestalt interactions; but we have made great progress in understanding how individual contour-elements can be detected. Studies by H. Keffer Hartline, Stephen Kuffler and Horace Barlow in the US and UK between the 1930s and the mid-1950s 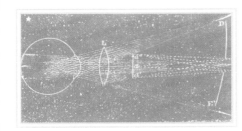 revealed cells in the retina that respond maximally to limited light or dark spots. At the John Hopkins and Harvard Medical School in the early 1960's, David Hubel and Torsten Wiesel discovered that cells of the primary visual cortex selectively respond to light-dark borders of different orientations. By their connection to a network of photoreceptors, such cells enhance our ability to sense the light-dark borders of certain orientations. We believe that the collective action of these orientation-sensitive cells are part of a mechanism that enables our visual system to define the edges of an object's form.

Increasingly, sophisticated psychophysical studies of visual perception are being combined with neurophysiological recordings of visual cell responses. Psychophysical studies of Edward Land, the inventor of the Polaroid camera, on the global mechanisms of color perception have inspired physiological investigations by Semir Zeki of University College, London, for instance, into how higher areas of visual cortex represent the correct colors of a scene in different light. Other scientists have trained monkeys to report their perceptions to simple visual stimuli directly while recording visual cell activity in the monkeys' brains. With the ability to perform psychophysical and neurophysiological studies together in the same higher primate, science is winding two strands of vision research together.

MAPPING THE FUTURE

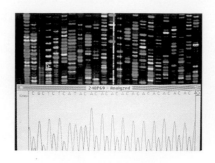

▲

DNA JOINED BY BASE PAIRING, AS SEEN HERE, IS THE WORK OF GENETICISTS WORKING ON THE HUMAN GENOME PROJECT.

1990

THE FIRST FOREIGN GENE is successfully introduced into maize by a biotechnology company in Cambridge, Mass. Genetic engineers hope they can introduce new genes into plants to increase their nutritional value and their tolerance to stress and disease.

91

TWO MOUNTAIN CLIMBERS scaling a glacier in the Austrian Alps stumble on the mummified body of a young man. Although the so-called Alpine Iceman has no wisdom teeth, he does display tattoos on his back and ankles and is fully equipped with a copper axe, bow and arrows, knapsack, and other implements. Carbon dating of the hay used as insulation in his shoes indicates he is 5,300 years old, placing him in the late stone age.

92

GENETICISTS FOR THE HUMAN GENOME PROJECT—the massive effort to map and identify each of the 100,000 human genes and work out the precise sequence of their 3 billion nucleotides—complete the study of 2 of our 46 chromosomes. Genes are found for a number of traits and illnesses, including predisposition to breast cancer, maturity-onset diabetes, and hypertension.

▶

AN AERIAL VIEW OF THE FERMILAB

COLLAPSE OF THE SOVIET UNION results in a Russian brain drain. The U.S. and Western Europe become common destinations for scientists fleeing suddenly unfunded research institutes unable to pay their salaries or buy modern equipment.

93

CONSTRUCTION OF THE SUPERCONDUCTING Supercollider is halted when Congress fails to approve the budget. Cost overruns and claims of mismanagement result in shutdown of the project that would have allowed physicists to re-create the conditions of the universe when it was one 100-quadrillionth of a second old—making it possible to discover such mysterious particles as the Higgs boson believed to give mass to all other matter.

PHYSICISTS AT FERMI National Accelerator Laboratory in Batavia, Illinois announce evidence for the top quark—the sixth and last of this fundamental subatomic particle's "flavors" to be confirmed (The others are bottom, up, down, strange, and charm.) Detected in only 12 of the 1 trillion particle collisions studied by the team of 440 physicists, the top quark is the most massive quark of all—nearly as heavy as a gold atom.

▲
LASER SCANNER
CAPTURING THREE-
DIMENSIONAL
IMAGES OF A FOSSIL

FOSSILS OF A NEW SPECIES of human ancestors, named *Australopithecus ramidus* (after the Afar word for root), are unearthed in Ethiopia. Living 4.4 million years ago, 800,000 years before Lucy and the other *Australopithecus afarensis*, they are thought to be the link between humans and apes.

1995...

Acknowledgments

We are grateful, above all, to the authors who, with their skill, brilliance and elegance, grace this collection with their thought-provoking accounts and anecdotes.

We are also grateful to everyone at Foca Company—Daniel H. Franck, Marius Muresanu, Melanie Roberts, Katie Andresen, and Tom Cannon, Jr. for producing a design that beautifully blends pictures and words.

Thanks to Kenneth Wright at Henry Holt and Company for his superb support and direction. Thanks also to Paula Kakalecik and Kevin Ohe at Henry Holt for their invaluable assistance.

The gracious contributions of time and mentoring by John J. Hanley, John J. Moeling, Marie Beaumonte, Edward Bell, and Kevin Gentzel of *Scientific American* are greatly appreciated.

Thank you, Bob Ubell, for your inspiration, and support. Thanks also to Elaine Cacciarelli, Mark Meade, Barbara Goldman and Aileen Acosta of Robert Ubell Associates.

Special thanks to Dana Desonie and Peter Pochna for their contributions to the chronologies, Paul Raeburn for his advice on the essays, Barbara Sullivan, Dick Hartzell, and Liesel Gibson for their talented work in copyediting, and Raissa Pockros for her assistance with transcribing.

Luis Gonzalez
Project Manager

PICTURE CREDITS

8 FPG International; **9-10** Tomo Narashima; **12** Photo Researchers; **16** (*right*) S.P. Thompson; **21** (*bottom*) Frans B.M. deWaal; **22** (*top left*) Peter Wellnhofer; (*middle right*) Photo Researchers; (*bottom right*) Sophie de Beuane; **25** (*right*) Simon Stoddart; **26** Alan Dressler; **34-35** Alfred T. Kamajian; **37** Alfred T. Kamajian; **42** Rothschild Estate; **44** Cambridge University Press; **45** Marcus E. Raichle; **48** (*inset*) Photo Researchers; **49** Bernard Sordat; **52** (*inset*) George H. Denton; **54** Markus Aellen; **55** Claude Lorius; **66** American Museum of Natural History; **70** Harry W. Green; **84** Art Resource; (*inset*) Patricia J. Wynne; **89** (*top*) Patricia H. Roberts; **94** (*inset*) GEOPIC, Earth Satellite Corporation; **97** (*bottom*) USDA Agricultural Research Service; **102** (*left*) NASA; **103** (*right*) Otto Hahn; **104** Library of The Swiss Federal Institute of Technology; **114** (*bottom*) NASA; **116** Mercury; **130** Axel Scherer; **134** (*top*) Photo Researchers; **107** Fred Andresen; **138** George Retseck; **144** Photo Researchers; **146** D.N. Harpp; **166** (*left*) NASA; **166** (*right*) Photo Researchers; **167** Fred J. Roberts; **168** NASA; **169** American Institute of Physics; **184** (*left*)-Photo Researchers; (*top*) NASA; **185** (*left*) Ian Worpole; (*right*) NASA; **186** Black Star; **192** Thomas R. Cech; **196** IBM Thomas J. Watson Research Center; **298** John Dewey Jones; **200** (*left*) Alfred T. Kamajian; **208** (*middle*) Art Resource; **212** NASA; **216** (*left*) Superconductor Technologies; (*middle*) Michelle Zawrothy; (*right*) Michael Goodman; **218** Oak Ridge National Laboratory; **220** (*top*) Photo Researchers; **221** (*bottom*) NASA; **222** Max Aguilera-Hellweg; **230** (*bottom*) Fred Andresen; **231** (*bottom*) Fred Andresen; **233** NASA; **234** (*top*) Photo Researchers; (*bottom*) Photo Researchers; **238** Arthur Tischchenko; **242** (*top*) Photo Researchers; (*bottom*) Fermi National Laboratory; **244** Rick Williams.

In addition, many of the engravings in the preceding pages first appeared in the pages of *Scientific American* from 1845 to the early 1900s.

A PEN RACK AND INKSTAND

COMBINED WITH A PERPETUAL CALENDAR, ILLUSTRATED

IN AN 1865 ISSUE OF *SCIENTIFIC AMERICAN*

Contributors

Daniel L. Alkon is Medical Director and Laboratory Chief at the National Institutes of Health in Bethesda. Dr. Alkon leads a multidisciplinary research program on the molecular and biophysical basis of memory in the brain.

Bruce A. Bolt is Professor Emeritus of Seismology at the University of California, Berkeley. Dr. Bolt is the author of several widely used textbooks on earthquake hazard reduction and geophysics.

Lewis M. Branscomb is Director of the Department of Science, Technology and Public Policy at John F. Kennedy School of Government at Harvard University.

Robert J. Cava is a Distinguished Member of the Technical Staff in the Solid State Chemistry Research Department at AT&T Bell Laboratories. He specializes in the discovery of new materials with potential for application in emerging electronic technologies.

Luca Cavalli-Sforza is Professor of Genetics at Stanford University. He is an expert in the areas of human population genetics and cultural evolution.

Thomas R. Cech, Nobel laureate in chemistry, is an Investigator at the Howard Hughes Medical Institute at the University of Colorado. He also teaches introductory chemistry at the university.

Patricia and Paul Churchland are at the Department of Philosophy at the University of California, San Diego. They are the authors of numerous works on computer intelligence.

Barry Commoner is Director of the Center for Biology of Natural Systems at Queens College, City University of New York. His most recent book is *Making Peace with the Planet*.

Ronald C. Davidson is Director of the Princeton Plasma Physics Laboratory and Professor of Astrophysical Sciences at Princeton University.

Frans B. M. deWaal is an ethologist and primatologist specializing in the social behavior of monkeys and apes. He is the author of *Chimpanzee Politics* and *Peacemaking among Primates*.

Sylvia A. Earle is at Deep Ocean Engineering in Oakland. Her work concerns oceans and their exploration.

Anne H. and Paul R. Ehrlich are Senior Research Associate and Bing Professor of Population, respectively, in the Department of Biological Sciences at Stanford University. Their most recent books are *Population Explosion* and *Healing the Planet*.

Carl Eisdorfer is Professor and Chairman of the Department of Psychiatry at the University of Miami. He is also Director of the Center of Adult Development and Aging as well as Director of the Center for Biopsychosocial Studies on AIDS.

Niles Eldredge is Curator of the Department of Invertebrates at the American Museum of Natural History. He is coauthor of the "punctuated equilibria" theory of evolution.

John Firor heads the Advanced Study Program at the National Center for Atmospheric Research. He is a former director of the Center and is the author of *Changing Atmosphere: A Global Challenge*.

Robert C. Gallo is Chief of the Laboratory of Tumor Cell Biology at the National Cancer Institute, National Institutes of Health. He is also Adjunct Professor of Biology at Johns Hopkins University and Adjunct Professor of Genetics at George Washington University.

CONTRIBUTORS

Howard Gardner is Director of Harvard Project Zero. He is currently investigating the application of intelligence theories in education.

John H. Gibbons is Science Advisor to the President of the United States.

Martin E. Hellman is Professor of Electrical Engineering at Stanford University. His work involves cryptography, technology and society, and the resolution of ethnic conflict.

Roald Hoffmann, Nobel laureate in chemistry, describes his work as "applied theoretical chemistry." He is currently at Cornell University.

J. Allan Hobson is Professor of Psychology at Harvard Medical School. He is also Psychiatrist and Director of the Neurophysiological Laboratory at Massachusetts Mental Health Center.

Piet Hut is Professor at the Institute for Advanced Study in Princeton. His work is in theoretical astrophysics and stellar dynamics.

Donald C. Johanson is President of the Institute of Human Origins. A leading paleoanthropologist, Dr. Johanson is most recognized for his announcement in 1978 of a new species, Australopithicus aferensis.

Mindy L. Kornhaber is a researcher at Harvard Project Zero. Her work concerns the application of intelligence theories by schools and educators.

David Lightfoot is Professor and Chair of the Department of Linguistics at the University of Maryland, College Park. His work involves syntactic theory, language acquisition by children, and language change from one generation to another.

Jonathan M. Mann is Center Director of the Francois-Xavier Bagnoud Center for Health and Human Rights at the Harvard School of Public Health. He is a leading proponent of the worldwide prevention and control of AIDS.

Gary T. Marx is Professor Emeritus at MIT and Professor of Sociology at the University of Colorado. He is currently engaged in research on the social aspects of information technology.

Ralph C. Merkle is a research scientist at the Xerox PARC. He is a co-creator of public-key cryptography.

Dorothy Nelkin is University Professor in the Department of Sociology and School of Law at New York University. She has written several books on the social and ethical aspects of science.

Kevin Padian is Associate Professor of Integrative Biology and Curator in the Museum of Paleontology at the University of California, Berkeley. Dr. Padian's current work includes the origins of major evolutionary adaptations, particularly the origin of flight and the beginning of the Age of Dinosaurs.

Stuart L. Pimm is Professor of Ecology at the University of Tennessee. He is a Pew Scholar in Conservation and the Environment.

Cyril Ponnamperuma is Professor of Chemistry and Director of the Laboratory of Chemical Evolution at the University of Maryland. Dr. Ponnamperuma has devoted his career to the study of the origin of life.

Peter H. Raven is Director of Missouri Botanical Garden. He is also Engelmann Professor of Botany at Washington University.

Richard Restak is Professor of Neurology at George Washington University. He is the author of ten books and numerous articles on the human brain.

Steven A. Rosenberg is Chief of Surgery at the National Cancer Institute. He has developed innovative immunotherapies and gene therapies for patients with cancer.

David C. Sabiston, Jr. is James B. Duke Professor of Surgery and Chief of Staff at Duke University Medical Center in Durham.

Jeremy A. Sabloff is the Charles K. Williams II Director of the University of Pennsylvania Museum of Archaeology and Anthropology. Dr. Sabloff is an internationally recognized authority on the ancient Maya.

Carl Sagan is David Duncan Professor of Astronomy and Space Sciences and Director of the Laboratory for Planetary Studies at Cornell University. Dr. Sagan has played a leading role in the American space program since its inception.

Joseph Silk is Professor of Astronomy and Physics at the University of California, Berkeley. He is a theoretical astrophysicist interested in span cosmology, particle astrophysics, and the formation of stars and galaxies.

Solomon H. Snyder is Director of the Department of Neuroscience at Johns Hopkins Medical School. Dr. Snyder carried out pioneering research identifying the opiate receptor and endorphins.

Howard Z. Streicher is at the NIH Chical Centre. His work concerns immunological responses to human viral infections.

Gerald Jay Sussman is Matsushita Professor of Electrical Engineering at MIT. His work concerns artificial intelligence.

Lester C. Thurow is Jerome and Dorothy Lemelson Professor of Management and Economics at the Sloan School of Management at MIT. He is Vice President of the American Economics Association.

Charles H. Townes, Nobel laureate in physics, is the inventor of the maser and coinventor of the laser. He is currently at the University of California, Berkeley.

Virginia Trimble is Professor of Physics at the University of California, Irvine and Visiting Professor of Astronomy at the University of Maryland. She has written extensively on supernovae, black holes, pulsars, and the large-scale structure of the universe.

Michael S. Turner is Professor of Physics at the University of Chicago and Staff Scientist at Fermilab National Accelerator Laboratory. His research concerns the earliest moments of the universe.

Robert A. Weinberg is Professor of Biology at MIT and is a Member of the Whitehead Institute for Biomedical Research. His research concerns the molecular basis of cancer.

Steven Weinberg, Nobel laureate in physics, holds the Josey Regental Chair of Science at the University of Texas. Dr. Weinberg is a recipient of the National Medal of Science.

James D. Weinrich is currently involved in AIDS research at the University of California, San Diego. He is the author of *Sexual Landscapes*.

Torsten Wiesel, Nobel laureate in medicine, is President of the Rockefeller University. His lifelong work has been the investigation of the neural basis of visual perception.

Garrison Wilkes is Professor of Biology at the University of Massachusetts. He is an expert in the area of origin and evolution of food crops.

Michael E. Wysession is Assistant Professor of Geophysics at Washington University in St. Louis. His work, using seismic waves from earthquakes to map out the interior of the earth, has led to a greater understanding of the dynamics of earth's mantle and core.

Index